Lesser Known Medicinal Plants

NIPA GENX ELECTRONIC RESOURCES & SOLUTIONS P. LTD.
New Delhi-110 034

About the Authors

Dr. Raviraja Shetty G, did his Ph.D in Horticulture from University of Agricultural Sciences, Bangalore, India. He worked on biodiversity, crop improvement, propagation and conservation of Rare, Endangered and Threatened (RET) medicinal plants of Western Ghats. He has handled DST-Fast track young scientist project entitled Exploration, Conservation and Utilization of RET Medicinal plants of Western Ghats. Dr. Shetty has received Environmentalist of the Year-2010 award by National Environmental Science Academy, New-Delhi, Young Scientist of the year 2014 award by Scientific and Environmental Research Institute, Kolkata, Scientist of year award 2020 by National Environmental Science Academy, New-Delhi and Best Faculty award by Education Expo TV, EET- CRS, Noida. Recently he has been conferred with "Biodiversity Conservation award" from Karnataka Biodiversity Board

He has chaired many sessions in national and international conferences in the field of Horticulture and medicinal plants. He has delivered lead talks and invited talks at various national and international conferences. He has wide international exposure and visited countries like Serbia, Portugal, France, Israel, Switzerland and Germany to present his research work and to attend international trainings. He has guided more than 20 M.Sc. and Ph. D students for their research as Major Advisor. To his credit he has more than 60 research papers published in national and international journals. He has handled various research projects funded by DST, NMPB, ICAR, , NHM and GOK. He is the member for Medicinal Plant Specialist Group and Species survival Group of International Union for Nature Conservation (IUCN). Presently he is working as Associate Professor and Head at Agricultural & Horticulture Research Station, Ullal, Mangalore under Keladi Shivappa Nayaka University of Agricultural & Horticultural Sciences, Shivamogga.

Tamanna Arif, pursuing her Ph.D at the University of Horticultural Sciences, Bagalkot, is an esteemed student who aspires to be a innovative scientist. She has done her under graduation and post graduation at the KSN University of Agriciltiral and Horticultural Sciences, Shivamogga. Her early achievements in a nutshell are; She has been a meritorious student and is going to be a recipient of Gold medal in her Post graduation. She has proactively participated and presented abstracts, posters and research work in various National and International Conferences. She has carried out her research work on 'Soilless cultivation of Ginger' and have been bagged 'Best Poster Award' for the research work by the University. She has also published about 20 articles in reputed magazines and journals and few articles of them have been awarded with various titles. Her achievements have been recognized and conferred with 'Emerging Scientist Award' at the International Scientist Conference on Engineering, Science and Medicine 2021. Currently, she is pursuing her first year of Doctoral degree at UHS, Bagalkot.

Lesser Known Medicinal Plants Cultivation and Utilization Techniques

Raviraja Shetty G.
Associate Professor (Horticulture)and Head
Agricultural & Horticultural Research Station
Kapikad, Ullal, Mangalore - 575020
(Keladi Shivappa Nayaka University of Agricultural
& Horticultural Sciences, Shivamogga), Karnataka

Tamanna Arif
College of Horticulture
UHS (B) Campus, Vignana Kendra
Bengaluru, Karnataka 560065
(University of Horticultural Sciences, Bagalkot)

NIPA GENX ELECTRONIC RESOURCES & SOLUTIONS P. LTD.
New Delhi-110 034

**NIPA GENX ELECTRONIC
RESOURCES & SOLUTIONS P. LTD.**

101,103, Vikas Surya Plaza, CU Block
L.S.C.Market, Pitam Pura, New Delhi-110 034
Ph : +91 11 27341616, 27341717, 27341718
E-mail:newindiapublishingagency@gmail.com
www: www.nipabooks.com

For customer assistance, please contact
Phone: + 91-11-27 34 17 17 Fax: + 91-11- 27 34 16 16
E-Mail: feedbacks@nipabooks.com

ISBN: 978-93-94490-32-1

Composed and Designed by NIPA.

Prologue

Interest in medicinal plants as a re-emerging health aid has been fuelled by the rising costs of prescription drugs in the maintenance of personal health and well-being. Herbal-based traditional medicine has become popular in developed countries in recent years, and its use is likely to increase in the coming years. This system has advantages over the modern medicine as it is prophylactic. Therefore, it would be a wise step if most of medicinal plants, which are lesser known to mankind but have immense potential, particularly those which are in high demand and rare should be cultivated and brought into limelight.

The basis for this textbook originally have been stemmed from the passion of both the authors in order to develop better methods of conveying and spreading the cultivation aspects of many crops and its uses by compiling the information available. As the world moves further into the advanced age, along with generating vast amount of information, there will be a greater need to access legacy plant materials with technology available. Thus this book helps in accessing the information on rare medicinal plants and thereby breaking down barriers of accessibility of limited information. The information provided have been compiled from various books, bulletins, write ups and journals.

Hope this book provides, the readers a comprehensive data and knowledge on cultivation and the uses of lesser known medicinal plants.

January, 2022

Raviraja Shetty G.
Tamanna Arif

Prospects, Challenges and Scope for Cultivation of Lesser Known Medicinal Plants

Among ancient civilizations, India has been known to be rich repository of medicinal plants. The forest in India is the principal repository of large number of medicinal plants. India has 15 Agro-Climatic zones and medicinal plants are distributed across all bio-geographic regions, diverse habitats and landscapes. Around 70% of India's medicinal plants are found in the tropical areas and around 30% in the temperate and alpine areas. World Health Organization (WHO) has estimated that approximately 80% of the world population still relies on traditional medicines, which are mostly plant-based drugs. About 6198 species of plants are estimated to be used for human and veterinary health care in the country, out of which about 2,700 plants species are reported to be used in the codified Indian Systems of Medicine *viz.* Ayurveda (1800 species), Siddha (500 species), Unani (400 species) and Amchi (300 species). In addition to their use in preparation of traditional medicines, the medicinal plants are being used in preparation of various pharmaceuticals and health products under the modern system of medicine. The global resurgence of interest in complementary and alternative systems in general and in Indian Systems of Medicine in particular is increasing the demand of Ayurvedic, Siddha and Unani (ASU) drugs, which use mostly medicinal plants as raw materials. Medicinal plants have so far largely been collected from wild resources. Moreover, the plant material collected from these sources is replete with the problems of adulteration and mis-identification. Therefore, cultivation of genuine, authentic variety of plants may be the only way to have raw material of required quality. The non-availability of proper techniques and authentic planting material are the main constraints in cultivation of these plants.

Herbal medicines encompass the combination of practices of indigenous systems of medicine and several therapeutic experiences of many previous generations which delivers valuable guidelines to the selection, preparation, and application of herbal formulation for the treatment, control, and management of a range of illnesses. Plant-based drugs are reported to be successfully used to cure skin diseases, tuberculosis, diabetes, jaundice, hypertension, mental disorders, cancer, AIDS, and many other infectious diseases. Countries with ancient civilizations like India, China, South America, and Egypt are still using several plant-based remedies for treating such ailments. According to World Health Organization (WHO), 60% of the world's population relies on herbal medicine and about 80% of the population in developing countries depends almost totally on it for their primary health care needs. Phytocompounds and also their chemical analogs have provided abundant clinically useful drugs in the treatment of chronic and acute diseases. And still research is continued to search for newer therapeutic agents from medicinal plants. The herbal industry shares about US$100 billion with decent growth potential worldwide. The WHO has stated that trade in medicinal plants, herbal raw materials, and herbal drugs is growing at annual growth rate of about 15%. The increasing popularity and acceptability of herbal medicine is belief that all natural products are safe, cheaper, and commonly available.

Out of nearly 3,00,000 species of higher plants available, only a small proportion have been investigated for medicinal properties, and a still smaller number yield well defined drugs. The same is the case with lower plants and with plants of the sea. Thus, the knowledge of plant constituents gained so far is still meager, considering the huge number of species available in the world. Approximately, only 10% of the organic constituents of plants are reported to be known and the remaining 90% are yet to be explored. A very small proportion of Indian medicinal plants are lower plants like lichens, ferns, algae, etc. The majority of medicinal plants are higher plants. The major families in which medicinal plants occur are *Fabaceae, Euphorbiaceae, asteraceae, Poaceae, Rubiaceae, Cucurbitaceae, apiaceae, Convolvulaceae, Malvaceae* and *Solanaceae.*

Drugs are derived from trees, shrubs and herbs and even from primitive kinds of plants which do not fall into the above categories. They are made from fruits (Senna, *Solanum viarum*, Datura, etc.), flowers *(Butea monosperma, Bauhinia variegata*), leaves (Senna, Datura, Periwinkle, Tylophora, etc.,) stems (Liquorice, Ginger, Dioscorea, Costus, Garlic) roots (Rauvolfia, Periwinkle, Ginseng, etc) seeds (Isabgol, Abrus, Nux vomica) and even bark (Cinchona).

Contents

List of Colour Plates

1

Introduction
The Scope and Importance

Plants have been used by man from prehistoric times to relieve from suffering and to cure various ailments. Primitive people, when injured in battle or when they had a fall or cut, they instinctively resorted to materials available at hand for staunching the flow of blood or for the relieving of pain and, by trial and error, they learnt that certain plants were more effective than others when used for these purposes. Man has also gained such knowledge from his observation of birds and animals that have been using plants for curing their ailments. Even today, we find that the domestic dog and cat, when they suffer from indigestion or other ailments, run to the field, chew some grasses or herbs and get cured. The folk medicines of almost all the countries of the world abound in medicinal plants and tribal people wherever they exist, rely chiefly on herbal medicine, even today.

Today, chemical and pharmaceutical investigations have added a great deal of status to the use of medicinal plants by revealing the presence of the active principles and their actions on human and animal systems. Investigations in the field of Pharmacognosy and Pharmacology have supplied valuable information on medicinal plants with regard to their availability, botanical properties, and method of cultivation, collection, storage, commerce and therapeutic uses. All these have contributed towards their acceptance in modern medicine and their inclusion in the pharmacopeias of civilized nations.

The indigenous systems of medicine practised in India are based mainly on the use of plants. Ancient medicine was not solely based on empiricism and this is evident from the fact that some medicinal plants which were used in ancient times still have their place in modern therapy. Thus for example, 'Ephedra' a plant used in china 4000 years ago is still mentioned in modern pharamacopoeias as the source of an important drug, ephedrine. The plant Sarpagandha (*Rauvolfia serpentine*) which was well known in India as a remedy for insanity is being used today for curing mental ailments. Quinine, another important antimalarial drug of modern medicine, was obtained from the cinchona tree.

The knowledge about the use of medicinal plants has been accrued through centuries and such plants are still valued even today, although synthetics, antibiotics, etc. have attained greater prominence in modern medicine. It is, however, a fact that these synthetics and antibiotics although they often show miraculous and often instantaneous results, prove harmful in the long run and that is why many synthetics and antibiotics have now gone out of use or have been specified to be prescribed strictly under medical supervision. In the case of most medicinal plants, however, no such cumulative derogatory effect has been recorded and that is why many of the medicines obtained from plants are still widely used today.

It is also true that lately, due to the rapid progress and spread of modern medicine, the popularity of herbal medicines has been gradually declining. But in a country where 80% of our population resides in small and remote villages, and most of them are economically backward, complicated prescriptions which are usually expensive and difficult to procure cannot be a thing of the masses.

Besides the above, the following are some of the reasons that make the cultivation of medicinal plants very important.

1. In nature there remains a wide variation among the plants with regard to their active principles. As only the best among them are used for cultivation, it enables us to obtain raw material of homogenous quality and high potency.
2. It is easy to grow and fulfill the commitment of large scale supply through cultivated sources rather than from natural sources, which mainly depend on nature for their regeneration and availability.
3. The growing pressure of population/urbanization and the development of roads to remote areas have resulted in deforestation and the loss of natural plant resources.
4. In many cases, as the important plant parts used are roots or the entire plant, these results in plant collectors engaging in destructive collection/ extractive methods, resulting in many plants becoming extinct or being listed as threatened.

   Despite the fact that our forests are reservoir of wild plants and that it serves a major resource-base for medicinal plants, perhaps, the importance of this has not been clearly spelled out with least importance been given to the long-term strategy for the conservation of biodiversity which in turn provide support to the communities which are solely dependent on forests for their livelihood.

2

Devil's Cotton *Abroma augusta* L.

Sterculaceae

Abroma augusta- the plant

Ayurvedic name	:	Pisachkarpas
Unani name	:	Ulatkambal
Hindi name	:	Ulatkambal
Trade name	:	Ulatkambal, Devil's cotton
Parts used	:	Root, root bark, stem, and leaves

Morphological characteristics

Abroma is a shrub or small tree that grows to a height of around 3-5 metres. The plant has horizontal and velvety branches that are 10-30 cm long, 6-18 cm wide, oblong or lanceolate, more or less cordate, delicately acuminate, membranous, whole, and three to five lobed with a petiole that is 1-12 cm long. The dorsal surface of the leaves is glabrous, while the ventral surface is hairy.

Floral characteristics

Abroma augusta- the flower

The plant has beautiful and intricate scarlet like flowers (5 cm in diameter) bearing purple colour. The inflorescence is a cyme. The sepals are persistent and lanceolate, measuring about 2.5 cm long. The petals are imbricate and will shortly fall off. Stamens are situated on a short staminal tube, and there are five staminodes. Each cell in the ovary is five lobed and pyramidal, with numerous ovules. The apex, the capsules are obpyramidal, membranous, finely pubescent, and truncate at the apex. Behind each carpel is a triangular wing. December and January are the fowering and fruiting months.

Distribution

The species is native to Indo-Malayan origin and is found growing all around the tropical belts of India, covering places of North-East and East Coast. The plant is also grown as an ornamental crop owing to its beautiful flowers.

Climate and soil

It is a plant found thriving well in the open conditions of tropics, grown at an elevations ranging up to 1,200 metres. The annual temperature congenial for its growth lies around 26-31°C, but it can also survive in temparatures varying from 18-36°C. Average annual precipitation of 1,500-3,000 mm, is said to be ideal for the growth, but tolerates 1,300-3,500 mm, and prolonged dry season is known to hamper the growth. Luxuriant growth is noted in hot and humid climate. The plant grows well in a variety of soils, like sandy loam to loam type.

Propagation material

Seeds are the most widely used source for the propagation. Black colured seeds, which are matured enough are to be collected in the months of December and January.

Nursery technique

Raising propagules

The seeds are to be pretreated before sowing. Nursey beds are well prepared and the pretreated seeds are sown during February to mid-March. The germination process starts few days after planting and is completed in about 12-15 days. The seedlings after the germination are transplanted into the polybags filled with loamy soil, sand, and FYM (farmyard manure) in equal ratio. Direct sowing of seeds is also practiced. Once the seedlings attain a height of about 20-25 cm they are ready for field planting during May-June.

Abroma augusta- the seeds

Propagule rate and pretreatment

A seed rate of 150–200 g per hectare is required for a plantation which generate 12000–14 000 seedlings. Seeds are treated with dilute sulphuric acid for six minutes to aid germination, after which they are thoroughly washed in running water to eliminate any acid residue. To prevent damping off, Bavistin 50 WP @ 0.20 percent and 0.25 percent–0.30 percent Rogor 30 EC should be applied to the soil in nursery beds and polybags.

Land preparation and fertilizer application

Light ploughing of the soil is done, and 30 cm x 30 cm x 30 cm pits are dug with a 1 m x 1 m spacing. When a base dose of FYM @ 10 tonnes/hectare is mixed with dried soil and filled in the pits, faster growth is achieved. The pit is fertilised with NPK (nitrogen, phosphorus, potassium) fertilisers at 30:40:20 kg/hectare.

Transplanting and optimum spacing

The best time to transplant seedlings into the pits is around May–June, when the seedlings have reached a height of roughly 20–25 cm. In the North-East, where rainfall is evenly spread throughout the year, rainfall may last until August. There should be enough moisture in the soil for the seedlings to establish themselves at the time of transplantation. In the field, optimal spacing is 1 m x 1 m, resulting in a crop stand of 10,000 plants per hectare.

Intercropping system

The plant can be cultivated as a single high-density crop because robust root growth is necessary for maximum output. Intercropping can obstruct root growth and should thus be avoided.

Interculture and maintenance practices

A top dressing of N @ 30 kg/hectare is applied 120 days after transplanting, following a basal dose of manure and NPK. During a 12-month period of soil working, dosages of NPK @ 30:40:20 kg/hectare are administered, followed by a treatment of N @ 30 kg/hectare after 18 months. The first intercultural manual weeding operation occurs 45 days after planting, followed by the second four months later. In the second year, intercultural weeding processes are carried out at the end of 12 and 18 months.

Irrigation practices

Under North-East India, the crop is cultivated in rain-fed circumstances. To maintain humidity, the crop should be irrigated on a regular basis. Water logging must be avoided as much as possible.

Disease and pest control

High moisture levels in the crop results in disease like damping off. Spraying Bavistin 50 WP @ 0.20 percent to prevent damping off and 0.25 percent–0.30 percent Rogor 30EC to prevent insect attack are two control strategies. On this crop, no further infections have been detected.

Crop maturity and harvesting

It takes roughly 24 months for the crop to mature. During the months of November and January, seeds are harvested. Post-flowering is the best time to harvest medicinally valuable portions. The roots are pulled out and washed with water before being harvested. Before digging out the roots, the soil should be adequately moist.

Post-harvest management

After harvesting, the bark should be sun-dried, packed in gunny bags, and stored in a cold, dry location.

Chemical constituents

Roots contain alkaloids abromine, abromasterol, and digitonide. The acceptable range of active constituents is 5.5%–5.7% of total constituents.

Yield

After two years, a yield of around 1.95 quintals of dry root bark is obtained per hectare.

Therapeutic uses

Abroma augusta root bark is a highly useful medicine used as an emmenagogus and uterine tonic, mostly for intra-uterine disease and other gynaecological disorders, especially menstruation disorders such as dysmenorrhoea, amenorrhoea, and gonorrhoea. Powder root is an abortifacient and anti-fertility herb. The leaves and stem have a sedative effect. A gonorrhoea infusion produced from fresh leaves and stems is effective. A infusion made from fresh leaves and stems is helpful in treating gonorrhoea.

3

Indian Aconite
Aconitum balfourii Stapf

Ranunculaceae

Inflorescence of *A. balfourii*

Ayurvedic name	:	Vatsanabh
Hindi name	:	Meetha vish
Trade name	:	Indian aconite, Meetha vish
Parts used	:	Tuberous root

Morphological characteristics

A. balfourii Stapf is an erect, glabrous shrub that grows upto 1.5 m in height. The plant has branched stems which are dull purplish brown in colur that reach about 1.2 m high. Lower leaves are long stalked while the upper ones are short. The plant has long heavy tubers with 7 to 12 cm in length. Roots are tuberous and fusiform.

Floral characteristics

The plants bears about 29 cm long inflorescence with numerous flowered racemes along with the yellow tomentum. The flowers are hermaphrodite and the mode of pollination is by bees. Sepals are helmet shaped, pubescent and blue in colur. The upper two petals are hooded and enveloped in the helmet, and the below two are glabrous. The stamens are many, the filaments are hispidulous, and the pollen is three-colpate. Carpals are yellowish, 3 to 5 in number and tomentose. Follicles range in size from 2 to 5, are sessile, and silky pubescent. Along the raphe, the seeds have a broad winged appearance. The months of September to November are ideal for flowering and fruiting in *A. balfourii.*

Classification on the basis of type of aconitine

Taking into account on the kind of aconite present in the species, Aconitum (Species which have been found in Uttarakhand region of India) have been classified into two classes: poisonous and non-poisonous Aconitum. In non-poisonous the species include *A. heterophyllum, A. laeve* and *A. routndifolium* whereas, in poisonous ten species have been included which are *A. chasmanthum, A. ferox, A. deinorrhizum, A. falconeri, A. balfourii, A. moschatum, A. violaceum,, A. spicatum, A. bisma and A. laciniatum.*

Distribution

Aconitum genus have been widely distributed in belts of Himalaya region and is endemic to alpine and subalpine belts of Indian Himalayan region where the elevation ranges from 3000-4300 AMSL. About 300 species of *Aconitum* have been found in the world out of of which 24 species are found in great Himalaya of India. Of all these, *Aconitum balfourii* Stapf is an important herb of the genus which is highly prized herb.

Soil and climate

Indian aconite is found growing at an elevation of above 2200 m altitude which is highly suitable for plant growth. Sandy loam soils which are slightly acidic (pH 5.1 to 5.5), rich in humus, are found suitable of cultivation of this crop. Plants have also performed well in calcareous soil. Partially shaded areas, thick soil and moist conditions provide a healthy environment for the plants.

Varieties

There are no known varities of this plant. When maintained at lower altitudes, strains taken from timberline populations demonstrate greater survival and response to vegetative propagation, growth, and production.

Propagation

The plant is usually propagation through seeds as well as tuber segments. At higher altitudes use of stem cuttings have also been found to be successful for plant multiplication.

Propagation study through tuber segments was carried out in alpine region in *Aconitum atrox*. It was attempted at higher altitude (Rawat *et al*., 1992). Similarly, the same study was attempted but at lower altitude and in their study, they successfully propagated the plant by treating tube segments with a combination of GA3, IBA, and Kinetin, a combination of any two of them or the use of any one of them for 48 h at room temperature. The study was helpful to overcome constraints in seed germination and seedling establishment.

Raising propagules

In nursery, seeds are planted at a soil depth of 0.5–0.7 cm in mist chambers or shadow rooms. Seeds can also be planted and mulched in open beds. Seed planting can be done in temperature-controlled chambers at moderate elevations (1800–2200 m) in October–November or March–April. Throughout May and June, sowing takes place in glass houses at lower elevations (for example, at 1550 m in Srinagar, Garhwal) and in open beds or hot houses in alpine locations. Tuber cuttings and seedlings should be planted at the same time. Tubers sprout in 6–45 days in the laboratory, while they sprout in 12–65 days in polyhouses with appropriate soil and temperature (15–20 °C). If tubers become available later in the season (October–November), instead of dividing into segments, whole tubers should be used at all altitudes.

Propagule rate and pretreatment

Nearly 2 kg of seeds are required to raise seedlings for 1 hectare of land or to sow 50 000 seedlings at a 45 cm x 45 cm spacing. Pretreatment of seedlings with GA_3 (gibberellic acid) (100 and 200 PPM [parts per million]) enhances germination. Sun drying of seeds before sowing, followed by mulching, is recommended for enhanced germination in open beds.

Land preparation and fertilizer application

During the winter, the field is ploughed thoroughly until a fine tilth is achieved. Weeds should be pulled and manure applied around 10–15 days prior to transplantation. Humus/compost/FYM (farmyard manure) can be added according to the requirements; for example, soils containing 1%–2% organic carbon will require FYM/compost at a rate of 25–30 quintals/hectare/year as a basal dose during land preparation. Forest litter applied at a rate of 100–150 quintals/hectare/year for the full crop season (5–7 years) has also been shown to produce good yields. However, there in no research on the use of inorganic fertilisers.

Transplanting and optimum spacing

During the months of March–April at lower and moderate altitudes, and May–August at alpine settings, seedlings are transplanted three months following the first true leaf initiation. Tuber segments can be planted straight in the field after it has been prepared. The recommended optimum spacing for tubers is 45 cm x 45 cm for better growth and development. Seedlings can also be transplanted at a 20–25 cm spacing at first, then thinned after two years. Around 50 000 seedlings are needed for a spacing of 45 cm x 45 cm, and about 200 000 seedlings are needed for a spacing of 20 cm x 25 cm. Because of desiccation or mechanical stress, the seedlings may die at a rate of 25–40% while transplantation.

Intercropping system

As a solitary crop, this plant is ideal. Intercropping with *Aconitum heterophyllum* has been attempted, however the death rate for *A. heterophyllum* when cultivated with *Aconitum balfourii* was as high as 80%.

Interculture and maintenance practices

It has been discovered that soil treated with high quantities of humus is ideal for the crop in order to achieve high yields. However, if there isn't enough leaf litter, FYM made from sheep or buffalo dung might be used instead. However, no seed or tuber production has been recorded at lower elevations, even in nursery beds treated with greater dosages of organic manure. Mulching and manuring is recommended during the winter season.

Irrigation practices

In cultivated fields, watering is not required during the monsoon season. Irrigation is dependent on selected site and soil condition. Until the seedlings are six months old, frequent watering at short intervals (two to seven days) is required at lower altitudes (1800–2200 m). The humidity of the soil should be maintained at all times.

Weed control

After the establishment of seedlings or the sprouting of tuber cuttings in the early growth period, a layer of broad leaf litter (up to 5-cm thick) is mulched. Weeding by hand is not required at this time. In the early stages of growth and during the winter season, constant weeding at a 20–30 day interval is essential.

Disease and pest control

Although insects may destroy the blossoms, there has been no evidence of a significant disease harming the plant. The crop is not thus treated with pesticides or insecticides.

Crop maturity and harvesting

The reproductive period of plants takes roughly five to seven years to complete. Plants grown from seedlings at lower altitudes, on the other hand, mature in three to five years. When grown from tubers, however, plants blooms in one to two years. Plants mature and after completing the reproductive phase achieve a high percentage of active contents. The length of time it takes for the reproductive phase to complete varies depending on altitude. Plants in alpine habitats normally finish their reproductive cycle in the last week of October or the first week of November, but plants at lower elevations usually finish in the first half of October. Plants grown from tuber segments have completed their reproductive phase by the end of the third year. During November–December, the most active components are detected. When the plant is six years old, the percentage of pseudoaconitine and aconitine is at its maximum especially during May–June,. During that period, plants can be picked to obtain a huge quantity of active compounds. For best yield and quality germplasm (seeds) for multiplication, plants should be collected in September at lower altitudes and October–November at higher elevations.

Post-harvest management

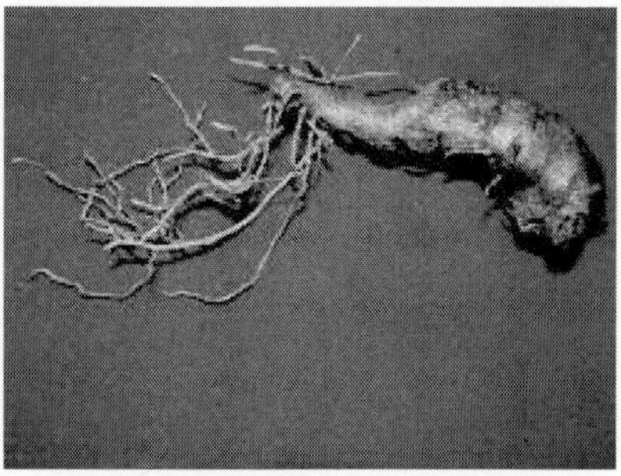

Tubers of *A. balfourii*

To harvest tubers, picking is done by simply digging the fields. The whole tuber is usually harvested. Small rhizomatic tuber segments (tubers with roots) can, however, be used for continued multiplication since they have a higher survival rate and growth percentage. The tubers with no rhizome are also sliced into 4–6 cm long slices. They are then dried at room temperature in the shade. Tubers should be stored in damp-proof containers in dark, dry, and preferably cool places for no more than six months after complete drying, when the moisture content in slices is less than 10%, after which time the quality begins to decline.

Chemical constituents

A.balfourii tubers contain minor amounts of aconitine, picroaconine, aconine, benzyl aconine, and hemonapelline, as well as a crystalline poisonous alkaloid termed pseudo-aconitine (0.4 percent–0.5 percent).

Yield

The predicted output of root is almost 450 kg dry weight based on the dry weight of tubers and total plants planted in 1 hectare of land, which is higher than the production indicated for natural sites. However, assuming a survival rate of 25% to 40% of transplanted seedlings/tubers, actual productivity has been estimated at 275 kg/hectare for seedling-raised crop and 345 kg/hectare for tuber-raised crop after the maturity of plants during the third year of cultivation through vegetative propagation. After three years of cultivation, seedling-grown plants produced a maximum of 302 kg/hectare.

Therapeutic uses

Bachnak roots include diaphoretic, diuretic, analgesic, febrifuge, anti-inflammatory, anti-rheumatic, anti-pyretic, and vermifuge effects. It is utilised to alleviate a variety of aches and pains. In excessive doses, it acts as a sedative, narcotic, and poisonous substance.

4

Angelica
Angelica glauca Edgew

Apiaceae

Ayurvedic name	:	Chorak
Hindi name	:	Chora, Choru, Gandrayan
Trade name	:	Gandrayan
English name	:	Angelica
Parts used	:	Roots/ Rhizome

Angelica glauca

Morphological characteristics

It is a glabrous aromatic perennial or biennial herb, 1-2 meter tall. Stem is hollow. Root is thick rhizomatous. Leaves are unipinnate, bipinnate, or tripinnate, large pinna is toothed, ovate or lanceolate.

Floral characteristics

Inflorescence is compound umbel with numerous rays. Flowers are white, yellow or purple in colour, bracteate; florets white or purple. Seeds are small in size and winged. Fruits/seeds are 1.25 cm by 0.6 cm in size.

Distribution

The plant is endangered in status, which is distributed in Western Himalaya from Kashmir to Uttrakhand, in alpine scrub and forest shades between 2700-3700 meters.

Climate and Soil

It requires cool and temperate climate. It can be cultivated between 2000-3000 meters above msl. It requires deep rich porous and moist soil with shady situations. For its ideal cultivation, the plenty of organic manure is required.

Propagation material

Seeds and rootstock spilts.

Fruits

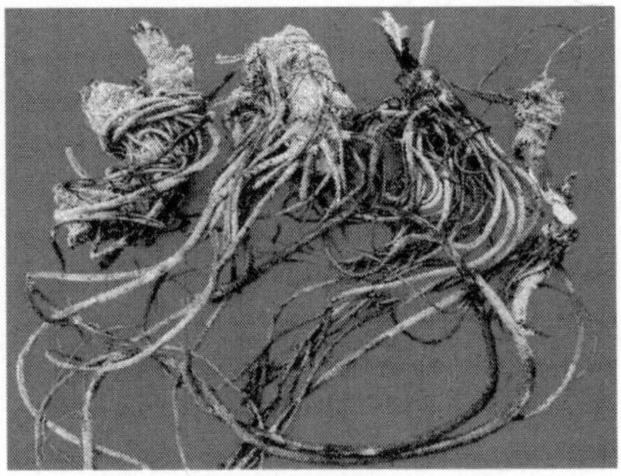

Rootstock

Nursery technique

Raising propagules

Seeds are sown immediately after harvesting during November and December inside polyhouse. Seed viability is very low and moist seeds have a better germination potential than dry seeds. Germination takes 25-40 days.

Propagule rate and pretreatment

Approximately, 50,000 plants or 6.2 kg seeds are required for its cultivation in one hectare land area.

Land preparation and fertilizer application

The field should be ploughed thoroughly followed by harrowing to bring the soil to a fine tilth and free from weeds. Seedlings are transplanted 45X45 cm apart in April and May. Apical portions of roots are transplanted during the rainy season at 45 cm apart. By this method, plants can be harvested within two years.

Green manuring

Sheep and goat manure is reported to be good for its cultivation. Approximately 15-20 tonnes of manure is required for one hectare of land initially, at the time of starting cultivation in lower altitudes. Manuring is done before planting. If required, manuring should be done after the completion of the vegetative growth phase during October or in the winter after two or three years of cultivation. At higher elevations, where forest litter is available, it enhances growth as well as survival and yield.

Transplanting the seedlings to main Field and optimum spacing

After four to six months growth of seedlings raised inside a greenhouse or in a small nursery, transplanting is done at the beginning of the rainy season. Raised beds are better for growth. If the site is moist or has good irrigation facilities, transplanting can be done during April and May.

Seedlings of *Angelica glauca*

Intercropping system

It requires similar climatic and edaphic conditions as *Saussurea costus,* so intercropping with this plant is beneficial.

Intercultural and maintenance practices

The intercultural operation like weeding / hoeing is carried out periodically as and when required.

Irrigation Practices

Irrigation twice a week during the dry season is required.

Weed Control

Weeding once a month and earthing every month during the rainy season and every two to three months during the dry season is essential.

Disease and pest control

No disease and pests have been reported.

Harvest management

Under cultivation, harvesting can be done within two to three years. Roots are harvested during September and October when seeds become partially mature.

Harvesting can be done after every two years, once the cultivation is well established and gives the maximum yield.

Post-harvest management

After harvesting the rhizomes, an apical portion is transplanted in the field for future crops. The remaining portion is washed with water to remove soil, and roots are cut into small pieces and put in partial shade for drying. After complete drying, roots are stored and packed in cloth bags.

Chemical constituent

Roots contain 1-1.5% volatile oil, valeric acid, angelic acid, lactones, sesquiterpenes, δ-α-cadinene, umbelliprenin, terpene alcohol and angelisine resin and essential oil.

Yield

At high altitudes of Garhwal, nearly 593-600 kg/ha yield is estimated under cultivation which is greater than all medicinal plants and has great potential as a cash crop. For a long time the collection of this species has been done by Bhutiyas from nature, and they still sell it as a condiment in valley villages to earn money. Nearly 15-25% *A. glauca* is sold in trade of medicinal plants by local people, especially from the Niti and Mana Valleys.

Therapeutic uses

A.glauca is used in dyspepsia, constipations, ulcer of palate, infantile atrophy, dysentery, menorrhagia and rinderpest. Roots/rhizomes are used as a drug for wounds and gastric pains. It has stomachic, carminative, stimulant, and sudorific properties. It is useful in anorexia, spasms, flatulent colic and bronchitis. The powdered root is administered in warm water for children's stomach ailments; it also checks vomiting. It is also used for flavouring confectionery and liquors. The leaves and stem act as a stimulant, cordial and used in stomach troubles.

5

Aloewood
Aquilaria malaccensis Lam. syn. *A. agallocha* Roxb.

Aquilariaceae (Thymelaeaceae)

Ayurvedic name	:	Agaru
Unani name	:	Ood Hindi /Agar
Hindi name	:	Agar
English name	:	Aloewood, Eaglewood, Agarwood
Trade name	:	Agar
Parts used	:	Fragrant Resinous Wood and Oil

Aquilaria malaccensis

Morphological characteristics

This plant is a large evergreen tree about 20 meters tall and 1.5–2.4 meters in girth with somewhat straight and fluted bole. Leaves are alternate 0.5-10 cm by 2-5 cm, oblong, lanceolate or elliptic, caudate, acuminate and glabrous with slender nerves. Venation is parallel. Petiole is 0.3-0.5 cm long. It is commercially used as fragrant and in prepation of drugs. The tree contains plenty of oleoresin and has irregular dark patches. The wood burns with a bright flame giving off pleasant smell.

Floral characteristics

Flowers are bisexual, white in colour, pedicillate, in both axillary and terminal umbellate cymes, shortly pedunculed, perianth, companulate, lobes 5 spreading and densely pilose. Pedicels is 0.5-0.8 cm long, slender. Perianth remains persistent in fruit and 1.3-1.5 cm long, silky densely villous, connate at the base. Stamens are 10, anthers 10 with subsessile disc. Ovary is subsessile, villous and two-celled. Stigma is large, subsessile. Fruit is capsular, 3-5 cm long, obovoid, pericarp coriaceous and densely tomentose. Seeds are ovoid with a long tail.

Distribution

The crop is found in parts of Bengal and North-Eastern States of India covering Assam, Meghalaya, Manipur, Mizoram, Arunachal Pradesh and Nagaland.

Climate and soil

This is a tropical tree that performs at its best in places receiving higher precipitation with humid condition. Temperatures ranging from 20^0C to 28^0C along with 80 % of relative humidity is suitable for the crop to cultivate. Sandy loam soils which are slightly acidic are preffered.

Propagation material

The crop is propagated through seeds.

Nursery technique

Raising propagules

Seeds tend to lose the viability very soon. Thus seeds are immediately collected when they mature i.e., during July-August and are sown within a week after its collection. Raising seedlings in poly-bag is preferred. Seed germinate with a success rate of more than 80%. About 4500 plants can be accommodated in per hectare area.

Planting in the field

Land preparation and fertilizer application

Before transplanting of seedlings, land should be thoroughly ploughed and harrowed to bring it up-to a good tilth. FYM @ 20 t/ha may be applied at the time of land preparation supplemented with NPK @ 60:60:40 may be applied in split doses. The fertilizer level is increased with age from 3rd year onwards.

Transplanting and optimum spacing

Seedlings when attain a height of 30-40 cm should be transplanted in the field during rainy season (April-June) at optimum spacing of 3X 3 meter.

Intercropping system

Annual or biennial medicinal herbs *viz Andrographis paniculata* (Kalmegh), *Withania somnifera* (Ashwagandh), *Rauwolfia serpentina* (Sarpagandha),

Bacopa monnieri (Bhrami), *Piper longum* (Pippali) *etc* may be cultivated as catch crops till the trees attain growth.

Inter-culture and maintenance practices

Spading and simultaneous weeding at 90 days after transplanting is required.

Irrigation practices

Grown as a rainfed plantation.

Weed control

Hand weeding is done after 90 days of transplanting, thereafter Gramoxone @ 0.5 kg/ha may be applied when necessary. Glycel @ 1.5 kg/ha may be applied to eradicate weeds.

Disease and pest control

Attack of *Heortia vitessoides* is observed during May-August. This causes defoliation of whole tree. Application of Thiodan @ 2 ml/lit at 15 days interval during infestation is found to control the pests effectively.

Harvest management

Agar-wood develops a peculiar, persisting strong odour because of infestation by a fungal identified as *Zeuzera conferta*, it penetrates the hard wood, through wounds, injury or borers. All attempts to induce artificial infestation have failed; it is a natural phenomeon. It develops black patches and stores resinous oil which is separated through distillation of the woody chips. This oil has high value in medicine and perfumery industry.

Crop maturity and harvesting

Time of harvesting depends on disease infestation in hard wood. Agar is regarded as a pathological product formed as result of infection. Black patches in the bark indicate occurrence of infection and can be used for harvesting hard wood to commercial use.

Post-harvest management

Wood chips or chips powdered mechanically without generating heat are soaked in water for 2-3 days and transferred to stainless steel vessel which is part of a distillation unit. The distillation is done for 30-36 hours. Oil and water is collected in a separator and stored. The oil and water ratio in the condenser is kept low on account of the high boiling point. Oil is stored in closed container preferable in Aluminum bottles.

Chemical constituents

Presence of the woody chips attains an essential oil from 0.8% to 2.2% in 8-50 years old plant. The wood contains hexadecanoic acid (25.0%), pentadecanoic acid (6.7%) and oleic acid.

6

Danti
Baliospermum montanum Muell. Arg

Euphorbiaceae

Ayurvedic name	:	Danti, Dravanti
Unani name	:	Danti
Hindi name	:	Danti
Trade name	:	Danti, Jangli jamalgota
Parts used	:	Roots, leaves and seeds

Baliospermum montanum - the plant

Morphological characteristics

The plant is a monoecious undershrub that grows up to 3.5 metres tall and has serrated leaves and stiff branches that emerge from the root. Lower branches feature big, typically broad, elliptical, three- to five-lobed leaves with rounded bases, while above branches have small, lanceolate leaves. Petioles range in length from 5 to 15 cm.

Baliospermum montanum - the fruit

Floral characteristics

The flowers are unisexual. The calyx of male flowers is globose, 2.5 mm long, four to five partite, glabrous or slightly pubescent, membranous, finely mottled with a disc of six glands, and glabrous or slightly pubescent. There are approximately 20 stamens in each flower. Sepals are ovate-lanceolate and glabrous, and the disc is roughly 2.5 mm in diameter in female flowers. Fruit is a three-lobed capsule that is around 8–13 mm long and hairy. Seeds contain an oily endosperm and are speckled and smooth. Flowers bloom in January–February, followed by fruit that mature a month later.

Distribution

The species grows in tropical and subtropical places with rainfall of around 1000 mm, such as the Himalayan foothills, Kashmir to the Khasi hills, and especially in the Vindhyas southward. In North and East Bengal, Chhota Nagpur, and peninsular India, it is fairly frequent.

Climate and soil

For optimal growth, the crop demands a humid environment. It is a shade-loving crop that may also be grown in open areas. It grows well in well-drained sandy loam soils with pH levels ranging from 6.5 to 7.5. It may, however, grow in soils with a pH of up to 8.2. Plants require a certain quantity of rainfall to thrive, and their growth is impeded in areas with low rainfall.

Propagation material

Seeds and terminal cuttings are the easiest means of propagation. Seeds collected during February to April are ideal. Seeds have an outer trilobed covering and collection of seeds can be started when the outer covering of fruits turn brown and dry.

Nursery technique

Raising propagules

Baliospermum montanum - seeds

The ideal time for the cultivation of crop is during the Kharif season (onset of monsoon). Last fortnight of June or during the onset of monsoons the crop is raised. Nursery is to be well prepared using sand, soil, and FYM (farmyard manure) in proportion of 2:1:1. Once the nursery is established, the seedlings can be transplanted in the main field or direct sowing of seeds in the rows is practised. The seeds are sown at a depth of 3 cm in the shade. Between two rows, a minimum gap of 8–10 cm and 3 cm between seeds is maintained. To compensate for any seed germination or mortality issues later, two seeds per hill may be sowed. The plants are regularly watered and fair amount of moisture has to be maintained. To enhance the plant population, cuttings, after treatment with rooting harmones, can also be directly planted in nursery beds or polybags. When cuttings are kept in shade rooms or mist chambers, they root more quickly. Cuttings-raised through stock, on the other hand, takes longer to reach transplantable size than seedlings.

Propagule rate and pretreatment

Around 7–8 kg of seeds are required to raise seedling stock on one hectare of land. The use of mechanical scarification to pre-treat seeds has resulted in increased seed germination rates. Germination is also improved by soaking seeds in water for two to three hours and then drying them in the shade before sowing. The germination process, on the other hand, takes roughly 25 days.

Land preparation and fertilizer application

The land should be given a deep disc ploughing followed by harrowing and levelling. The field is divided into appropriate beds, usually 4–5 m × 10 m in size. One irrigation channel is kept between the two rows of beds. A basal dose of 10 tonnes/hectare of FYM is applied at the time of field preparation. In addition, NPK (nitrogen, phosphorus, and potassium) @ 75:50:25 kg/hectare may also be applied as basal dose at the time of transplantation.

Transplanting and optimum spacing

Ideal time for transplanting of seedlings is after 40 days of sowing. As said, cuttings when used as the propagating material, it takes longer time for plants to develp. The optimum spacing of 1.5 m × 1.2 m is recommended for crop cultivation. One hectare of land accommodates about 5500 plants.

Intercropping system

Since the crop has a dense growth, *Baliospermum* is usually grown as a monocrop. However, in fruit orchards and beneath the trees, the crop can also be raised as a catch crop.

Interculture and maintenance practices

Fertilizer is applied once the field is prepared and when the plants are transplanted. In general, half of the N, as well as the full doses of P and K, are administered at planting, with the remaining half of the N dose, 75 kg/hectare, given 60 days later. At 30 days following transplantation, an intercultural operation is required. Weeding cum hoeing are done every 30–45 days.

Irrigation practices

Irrigation should be done immediately after transplanting for establishment of plants, followed by irrigation every seven days during the summer season and every 20 days during the winter season.

Weed control

After 30–35 days of transplanting, the first weeding should be done manually. Weeding should be done every 30 days to four months to keep the field free of weeds. After five months, there is sufficient vegetative growth that in turn prevent weeds from emerging in the field.

Disease and pest control

There have been no reports of major insect pests or diseases. When the crop is around six months old, ants are found attacking the plant. The ants are drawn to the aqueous sap that leaks near the bloom. On the crop, no chemical insecticides should be utilised.

Crop maturity and harvesting

It takes ten months for the crop to reach full maturity. Flowers bloom in January–February, followed by fruits a month later. Fruit develop green and matures to a brownish black colour. Plants are dug up individually, and their roots and seeds are dried in the shade.

Baliospermum montanum –the roots

Post-harvest management

The produce is dried under shade and stored in dry cool areas in gunny bags.

Chemical constituents

The seeds contain axillarenic acid, whereas the roots include 12-deoxy-5-hydroxyphorbol-13-myristate, 13-palmitate, 12 deoxyphorbol13-palmitate, baliospermin, and montanin.

Yield

A dry root yield of 2 tonnes/hectare is obtained in a 10-month-duration crop.

Therapeutic uses

Baliospermum roots and leaves have cathartic, pungent, thermogenic, purgative, anthelmintic, and diuretic properties. Dropsy, anascara, and jaundice are all treated using the roots. Asthma is treated using a decoction of leaves. Seeds are purgative, stimulating, and rubifacient when used externally. The oil extracted from the seeds is hydrogogue, cathartic, and used to treat rheumatism.

7

Brazil Wood, *Caesalpinia sappan* L.

Fabaceae

Ayurvedic name	:	Patangah
Unani name	:	Pattang
Hindi name	:	Patang
Trade name	:	Patang, Brazil wood
Parts used	:	Wood, leaves, pods and seeds

Caesalpinia sappan - the tree

Morphological characteristics

Caesalpinia sappan is a thorny, small to medium-sized plant with prominent spines on the stem and leaf rachis that grows up to 10 metres tall. Branchlets are lenticellate, dull, and usually have paired, recurved stipular prickles. With 9–14 pairs of pinnae, the leaves are big, hairy, or glabrate.

Caesalpinia sappan - the flower

Floral characteristics

The inflorescence is a supra-axillary and terminal raceme with golden yellow flowers that creates a huge panicle. The fruit is an elliptical pod that ranges in colour from brown to black. The peak flowering and fruiting season is from July to September, with November being a rare exception.

Distribution

The crop is native to the Indo-Malayan region. The species can be found throughout the Indian peninsula. It is also produced as a plantation crop in all of the Southern states, as well as in Odisha and Bengal to the north.

Climate and soil

The plant is xerophytic, meaning it thrives in dry, hot climates. It grows best in subtropical and tropical environments. It may be cultivated in a variety of soil types. Its growth is best suited to red loamy soils rich in humus nutrients.

Propagation material

Planting is usually done with stock that has been raised from seeds. Fruits appear 5–15 days after flowering and mature in three months, i.e. between October and December. In November and December, seeds for the nursery are collected. The seeds should be soaked in GA at 300 ppm for 6 minutes after being treated with strong sulphuric acid. They should then be sown in poly bags. During the months of April and May, seedlings are raised. After 75 to 90 days, the plants are transplanted to the main field. Each plant should be separated by 4 metres. Top soil, dry cow dung, and sand are used to fill them. The crop is irrigated bi-weekly throughout the dry season.

Nursery technique

Raising propagules

The months of April and May are ideal for cultivating seedlings. The polybags are used to raise a nursery. Sulphuric acid-treated seeds are sown in 1:1:1 ratio in polythene bags containing sand, FYM (farmyard manure), and red laterite soil. During the months of June and July, the seedlings are ready for transplanting in the main field.

Seed rate and pretreatment

For raising seedlings for planting on 1 hectare of land, 600–800 g seeds may be required. For six minutes, the seeds should be treated with diluted H_2SO_4. The radical and the plumule may be damaged by concentrated H_2SO_4 and soaking in acid for an extended period of time. Thus seeds should be washed thoroughly under running water. To enhance quick germination, the seeds should be shade-dried for 24 hours and then immersed in 300 ppm GA_3 (gibberellic acid) for six hours.

Caesalpinia sappan - the seeds

Land preparation and fertilizer application

Ploughing and fine tilthing of the soil are required. Then, at a spacing of 4 m × 4 m, pits of 45 cm × 45 cm × 45 cm should be dug and filled with organic mixture. As a basal dose before planting, 6–8 kg FYM combined with 100:50:75 g NPK (nitrogen, phosphorous, potassium) per plant is excellent.

Intercropping system

Other crops cannot be grown in the shade where this species is grown. As a result, the plant is preferred as a single crop. In the early years, however, some annuals may be planted between the rows of Caesalpinia.

Transplanting and optimum spacing

After reaching a height of 30–40 cm, the 30–45 day old seedlings are transplanted in the field pits immediately after the commencement of monsoon rains. If there isn't any rain, water the field as soon as possible after transplanting. 625 seedlings at a spacing of 4 m × 4 m are accommodated in 1 hectare of land.

Interculture and maintenance practices

In the second year, apply 150:60:160 g NPK per plant, and in the third year, apply 200:100:150 g NPK per plant. Manuring and watering are done twice, once in May–June and again in September–October, by opening rings around the stem base at a distance of 30 cm. Gap filling for seedlings, as well as staking of seedlings, can be done 30 days following planting. The entire field should be weeded during the second week of September, followed by fertiliser application. Once every four months, manual weeding around the basin is advised.

Irrigation practices

Light watering should be done once daily for the first 15 days after transplanting, then weekly or fortnightly depending on the season and water requirements. Irrigation should be done in a furrow, however drip irrigation can also be used.

Disease and pest control

The Pod Borer bores into the pod and consumes the seeds. It can be controlled by spraying 0.2 percent nuvacron at weekly intervals during the fruiting stage. Termites eat the roots and trunk of the plant, causing it to dry out. Chlorpyriphos (2 ml/litre dilution in water) should be sprayed on the damaged plants.

Crop maturity and harvesting

Flowering occurs after the plant reaches the age of two to three years, and the pods and seeds can be harvested in October–November. After about 10–15 years after planting, the wood can be harvested, during the months of October and November,

Chemical constituents

The pod shell has 4% tannins, and the plant's bark has 1.8 percent tannins, as well as aromatic chemicals including brazilin, sappanchalcone, ceasalpin-J, caesalpin-P, and protosappanin A and B.

Yield

It is possible to get a yield of 2000–2500 kg pods per hectare per year, yielding 200–250 kg dry seeds per hectare per year. There are no estimates for wood production.

Therapeutic uses

Caesalpinia is primarily an emmenagogue, haemostatic, and antiinflammatory herb. It is one of the constituents in the indigenous medicine 'lucol,' which is used to treat non-specific leucorrhoea. In traditional medicine, the heartwood is used to cure contusions and thrombosis. Rheumatism, haemorrhages, and wounds are treated with the wood paste. Because it is resistant to sun rays, light, heat, and water, the heartwood provides a red dye that is utilised for cosmetic purposes. Anti-cancer and anti-diarrhoeal effects are found in the plant. Antibacterial and antifungal properties are found in the oil extracted from the leaves.

8

Perfumed Cherry
Callicarpa macrophylla Vahl.

Verbenaceae

Ayurvedic name	:	Priyangu, Priyaka
Unani name	:	Habb-ul-Mihlb
Hindi name	:	Dahiya
English name	:	Perfumed Cherry
Trade name	:	Dahiya & Sumali
Parts used	:	Leaves, fruits and roots.

Callicarpa macrophylla – the plant

Morphological characteristics

The plant is an erect undershrub, 1.5–2.5 m tall. Leaves are elliptic-oblong to lanceolate or ovate to ovate–lanceolate, 12–25 × 5–11 cm, acute or acuminate at apex, acute or cuneate at base, glabrescent, crenate-dentate; densely stellate–tomentose beneath; petiole 4–12 mm long; Fruits globose, drupes or berries, white to yellowish-brown with or without fruit stalk, fresh being succulent, 1–3 mm in dia. Intact fruits are smooth and brownish in colour and exhibit centrally located bilocular carpel and 4 nutlets each embedded with a yellowish white seed, in a transversely cut surface of a fruit;

Floral characteristics

Flowers appear in July-November are small, 0.5 cm long, pink in colour and crowded in axillary peduncled globose cymes, solitary, pedunculate or sessile. Flowers are bisexual, actinomorphic and fragrant. Calyx 4 or rarely 5 toothed, pubescent, corolla infundibular, 4 or rarely 5 lobed. Stamens are 4, exserted, anthers oblong 2 loculed, dorsifixed. Ovary is superior, 4 lobed, 4 loculed, 1 ovule in each locule, style short, stigma 2 lobed. Flowering & fruiting occurs during June–Dec. Fruits when tasted at first are somewhat sweet, later bitterish; fruit has fragrant odour specially after slight bruising the fruit. The fruits of *Callicarpa macrophylla* are edible.

Distribution

Plants of this species grow in open and secondary forests along the upper Gangetic plains, West Bengal Plains, Eastern and Western Himalayas, Kashmir, Assam, and northern Andhra Pradesh up to an altitude of 1800 meters. The species grows in other places worldwide, such as Nepal, Indonesia, China and South Korea.

Climate and soil

The plant grows in dry tropical forests including shaded places of ravines and scrub forests.

Propagation material

The crop is propagated through seeds.

Raising propagules

In the summer months, from June to July, seeds are sown at a rate of 5-10 kg/ha. Success of germination is about 50%. Propagation is also accomplished by stem-cuttings. The row-to-row distance should be 50 cm and plant-to-plant is 25 cm. Seeds commence germination after 30 to 45 days and stem cuttings sprout after 8-10 days.

Land preparation

The field should be ploughed two or three times in order to have the fine tilth to facilitate planting and sprouting of stem cuttings. The FYM should be mixed with the soil at a rate of 20-25 t/ha. The NPK should be applied at 40:20:10 kg/ha as a basal dressing and the same amount as a top dressing after establishment of plant population.

Intercropping

A crop can be raised as a sole crop for commercial purposes or in combination with other plants such as rosemary or garden sage.

Intercultural Operations

It is done after 30-45 days and second at 60-80 days.

Irrigation

The crop is cultivated as a rainfed crop.

Weed control

During the initial growth phase of the crop, weeds may hinder the crop's growth. Therefore, constant weeding and hoeing should be performed.

Disease and pest control

No visible symptoms of any pest and diseases have been recorded.

Crop Maturity and harvesting

Crop requires about 10 months to mature. The mature fruits are plucked during second year in November-December. Following the first year, the yield increases in the subsequent phases.

Post-harvest management

After plucking of fruits, these are dried in shade and kept in air tight vessels.

Chemical constituents

In fact, the leaf and fruit oils are both rich in selinene derivatives, with the fruit oil containing 41.6% beta-selinene and 6% alpha-selinene. Dendrolasin, another natural furanoids sesquiterpenoid, is also found in the leaf and fruit oils. The content of luteolin increased gradually with the growth of plants and reached the peak at the end of growth period. Total flavonoid accumulation of plant changed along with the growth of the plants, i.e. the contents increased gradually in the trunk and root, decreased in leaves.

Leaves contain ursolic acid, oleanolic acid; bark contains methyl betulinate and baurenol. Heart wood contains oleanolic acid and B-sitosterol.

Yield

One hectare plantation gives yield of 630 kg of dried-fruits and 1660-1680 kg of dry roots.

Therapeutic uses

Fruits are used in the therapeutics with symptoms likeburning sensation in the body, fever, vomiting, blood disorders, vertigo, nervous system and rheumatic diseases while inflorescence are used in the therapeutics of burning sensation in the body, fever, blood pitta diseases, amoebic dysentery and hyperhydrosis. The

Bark is used in the treatment of rheumatism and gonorrhea. Leaves are used for the treatment of stomach disorders In Ayurvedic literature, Priyangu is described as cooling, refrigerant, deodorant and antipyretic, tones up the digestive system, checks excessive perspiration, disinfects intestines, controls diarrhoea and allays burning sensation during fevers. Leaves are applied in rheumatic joints. Oil from root is aromatic.

9

Gandrayan
Cinnamomum camphora Nees & Eberm

Zingiberaceae

Sanskrit : Karpurah

Hindi name : Karpura

Trade name : Gandrayan

Parts used : Bark

Syn : *Laurus camphora* (Linn.) *Camphora officinarum* Nees

Camphor tree

Camphor (*Cinnamomum camphora*) known under the names camphor tree, camphorwood or camphor laurel of the family Lauraceae is obtained from the bark of camphor tree that is a principle source of camphor. It has its use throughout the globe and along the centuries as an ailment for various remedies that help in treating issues like irritations, pains, inflammation, congestion, etc. In addition to this *Cinnamomum camphora* also owns suppressive and anti mutagenic effect that is useful in treatment of deadly disease like human cancer cells retaining the healthy cells. Indian religious system is well versed with the utilization of Camphor in rituals, however the camphor used here is made synthetically.

The botanical name *Cinnamomum camphora* is derived from Greek language from the word 'kinnamomon' which refers to spice. This herb has its earliest record that is used by Egyptians for embalming purposes. Along with this it also finds its use in Sanskrit literature also. Camphor tree is a fragrant evergreen tree that is native to China, India, Mongolia, Japan and Taiwan and a variety of it is grown in Southern United States; especially in Florida. The species Camphora, referred to as camphor or karpura, is an important chemical ingredient present in the oil that found in the tissues of tree.

Morphological characteristics

It is a tall tree that grows up to 100 feet. The trunk is 8 m long and 2 m wide and bark is yellowish brown colored rough on the outside and smoother inside with vertical fissures. All parts of the tree have peculiar fragrance of camphor. Leaves are fragrant, leathery and are yellowish light to dark green with glossy veins with varied shape.

Floral characteristics

Yellowish white flowers occur in cluster and inflorescence is terminal panicle. Fruits are dark green colored that later turn black. Seeds are small and are aromatic. By the end of November, the dark blue berries fruit ripen.

Camphor is formed in oil cells distributed in all parts of tree. These cells begin to form early in growth of plant and are filled with a yellow oil from which camphor is slowly deposited. Oil becomes colourless progressively and volatile and irregular masses of camphor appear. Formation of camphor is brought about by an enzyme present in growing parts of tree, particularly in tissue within cambium region. Each layer of wood, as it is formed, is enriched by camphor.

Distribution

Camphor tree is a native of Japan, China and Taiwan. It is distributed over tropical and sub-tropical countries of Asia, Africa and South America for its woods from which camphor oil is derived. In India, plant is cultivated in Dehra Dun, Calcutta, Bangalore and Nilgiri hills.

Soil and climate

The native range of *C. camphora* is in the Asian eastern monsoon zone characterized by predominantly summer rains and winter temperatures rarely below freezing, but has now become abundantly naturalized in a wide variety of climates throughout the world. In areas where the species has naturalized successfully the climate is subtropical. Being native to the eastern monsoon region, climatic requirements for optimum growth of *C. camphora* include sub-tropical warm, moist summers and mild winters. These conditions are found in temperate or subtropical regions at altitudes of 50-1000 m and in tropical regions between 1000-2000 m. The best growth occurs with mean annual rainfall greater than 1000 mm and uniform rainfall distribution. The preferred mean annual temperature is above 16°C, but it can survive lower temperatures down to -9°C. Seedlings may suffer slight frost damage in some years *C. camphora* is a light-demanding species that readily colonizes exposed fertile soils. As a seedling it survives light shade, but once above 2-3 m tall it requires full overhead light. Site conditions and crop stand density influence the content of camphor and oil; the highest quantity of camphor has been found in trees growing on sandy loams and the lowest content on clay. In subtropical eastern Australia it can withstand dry spring conditions once established. Although adapted to establish in open ground, it can tolerate some shade and form single-dominant stands which excludes other species for many years.

It grows best on elevated free draining, acid, krasnozem (red ferrosol) soils in a subtropical climate where annual rainfall is 1000-2000 mm and frosts are infrequent. The soils dominated low pH and high levels of aluminium and iron oxides are suitable. It is also found on other soil types and lowland areas, although it is slower growing there. It is not found close to the sea nor does it colonize eucalypt forests. *C. camphora* prefers light/medium sandy loam to sandy clay loam, and freshly drained fertile soils. The pH value ranges from acid to neutral, but the species can tolerate saline soils <0.2%). It does not grow well on heavy-textured and infertile soils with impeded drainage, but does well on moderately heavy soils. It grows well on either flat or sloping sites

Propagation

Propagation is completely by seeds, produced in large numbers of fruits (up to 100,000 in mature trees) which are consumed by a variety of birds, especially pigeons in eastern Australia, and dispersed over their territory during the southern autumn and early winter. Water dispersal also occurs, fruits floating for up to 20 days in water with no effect on germination. Seed longevity varies depending on water content and environment but appears to be < 3 years. In an experimental study it is stated that viability of seed stored in the laboratory was approximately 13% after 2 years and 0% after 3 years. Germination from surface-sown seeds was virtually nil under natural rainfall and 1% remained viable after 12 months and nil in the laboratory after 8 months when dry.

Germination appears to be inhibited while the seed remains in contact with the flesh of ripe fruits, which appear to release inhibitory substances. Germination commences at 8-12 weeks in the glasshouse and extends over 22 weeks. Germination in the laboratory was enhanced by removal of the seed coat. Unwashed seeds of mature, green (unripened) fruits demonstrated earlier germination (17% after 8 weeks compared to 2-7% for seeds of ripe fruits) and more uniform germination in the glasshouse. Water immersion and transport does not impair germination, as indicated by soaking of excised seeds and whole fruits for up to 40 days and 20 days, respectively, in water. Germination occurs in spring in the field, presumably after winter stratification and dilution of remnants of inhibitory substances in the fruit and seed coat.

Cultivation

Nursery is maintained weed free. After 12-16 months seedlings are transplanted in main field in 60 cm^3 pits, 2-3 m apart. Application of organic manures and inorganic fertilizers has proved beneficial. Plants are trimmed to a height of 1.5-2 m and maintained as bushes to facilitate picking of leaves.

Pests and diseases

Disease organisms causing damage in the native range of *C. camphora* include *Phellinus noxius* (brown root disease) and *Botryosphaeria dothidea* canker. Termites have been recorded in introduced trees in Egypt, and the blue triangle butterfly in Australia, and avocado brown mite (*Oligonchus punicae*). *Plumosa emarginata* and *Glomerella cingulate* leaf blight have been recorded in India, otherwise there are no recorded pests or diseases where the species has been introduced.

Extraction and utilisation

Camphor oil is obtained by distilling leaves, twigs and wood. Crude camphor oil is separated in various fractions as white, brown and blue camphor oils. White camphor oil is generally not used as such in perfumes, but it serves as a starting material for production of a number of perfumery chemicals such as cineole, terpineol, menthol and thymol. There is little difference in total yield of camphor when two or four pickings are taken in a year. Tender leaves and plants grown in open contain more camphor. Yield of camphor and camphor oil is 50-80 kg/ha which varies widely with part used and geographical location.

Chemical composition

Camphor powder

The leaf of *Cinnamomum camphora* is composed chiefly of camphor, as the main element along with cineol, linalool, eugenol, limonene, safrole, a-pinene, camphene and some other components. The oil obtained contains eugenol that has harsh odour and the leaf oil is also natural source of linalool (94.9%) also citronellal (2.4%). The fragrance imparting constituents are Menthol, Thymol, Phenol, Salicylic acid and Naphthol etc. Its bark chiefly contains as cinnamaldehyde that imparts it the very peculiar odor and flavor. Safrole found in the plant is thought to be carcinogenic.

Different forms of camphor with therapeutic uses

- Paste - Paste is useful for treating the skin disorders that helps in relieving from pain. It also vaso dilates the area it is applied on.
- Powder - It serves as pain killer; it regulates the blood circulation, blood purification and build up the heart muscles. It also dehydrates the body. It cures indigestion, enhance salivary secretion, maintains oral hygiene. It is a good fungicidal and bactericidal agent. It also acts on the skin related problems by purifying the blood.
- Decoction - It treats cough and helps in dislodging of excess of mucus of the respiratory tract. It also discharges the poisons in the body. It also prevents urinary tract infections and provides strength to it.

10

Tree Turmeric
Coscinium fenestratum (Gaertn.) Colebr.

Menispermaceae

English name	:	False calumba
Hindi name	:	Jhar-i-haldi
Trade name	:	Tree turmeric, Pitchandan
Parts used	:	Woody stem, root

Coscinium fenestratum – the plant

Synonyms

Menispermum fenestratum Gaertn., *Coscinium peltatum* Merr., *C. wallichianum* Miers, *C. maingayi* Pierre, *C. usitatum* Pierre, *C. blumeanum* Miers var. *epeltatum* Boerl., *C.wightianum* Miers, *C.miosepalum* Diels, *C.fenestratum* var. *macrophyllum* Yamamoto, *C. fenestratum* var. *ovalifolium* Yamamoto, *Pereiria medica.*

In India, vernacularly it is known as Haldigach (Bengali), Tree turmeric, False calumba, Columbo weed (English), Jhar-I-haldi (Hindi), Maradashina, Maramanjali (Kannada), Maramanjal, Manjavalli (Malayalam), Jhade-halade, Venivel (Marathi), Darvi, *daruharidra*, Pitadaru (Sanskrit), Venivel, Bangwellgetta (Sinhalese), Maramanjal, Manjalkodi, Atturam, Kadari, Udaravi, Pasamantram, Imalam (Tamil), Manu pasupu (Telugu).

Morphological characteristics

It is a huge dioecious tree that can reach a height of 10 metres and has yellow wood and sap. The slices of stem and root are stiff and woody. The colour of the wood is yellowish-brown on the outside and yellow on the inside. The substance is found in huge woody, cylindrical, straight sections with a diameter of up to 10 cm; branchlets are terete, brown tomentose, and subsequently glabrescent with disciform petiole-scars. Leaves simple, alternate, exstipulate, broadly ovate, rounded, truncate or shallowly cordate at base, acuminate at apex, 10-32x8-22 cm, glabrescent above, hoary yellowish-white tomentellous beneath, thinly coriaceous; main nerves 5-7, palmate, with 2 pairs of distal lateral nerves; midrib and other main nerves sunken, whitish tomentose beneath above basal margin on the lamina of leaf; stipules absent.

Floral characteristics

Inflorescence is globose heads on 1-3 cm long peduncles, of 5-11 cm long racemes, 6-7 mm broad, supra-axillary or on old leafless stems; bracts subulate, villous, tightly pressed on the calyx; peduncle bracts tiny, 4-5 mm long. Sepals are 9 in number, in three whorls, imbricate, densely sericeous-pilose; petals absent; flowers unisexual, tiny, yellowish or whitish; sepals 9, in three whorls, imbricate, densely sericeous-pilose; Sepals broadly elliptic to obovate, 1.5-2 mm long, densely sericeous outside, glabrous inside, yellow; outer ones 3-6, broadly elliptic, 1-1.5 mm long; inner ones 3-6, spreading, 1.5-2 mm long, yellow. Male flowers: sessile or shortly pedicellate, ca. 1 mm long; sepals broadly elliptic to obovate, 1.5-2 mm long, densely Anthers tiny, oval, adnate, outer ones one-celled, inner ones two-celled; pollen grains oblate-spheroidal or rarely spheroidal; filaments. 1 mm long; anthers small, oval, adnate, outer ones one-celled, inner ones two-celled; pollen grains oblate-spheroidal or rarely spheroidal.

Female flowers: 3-6 free, subglobose carpels with slender subulate recurved or filiform styles, staminodes 6. Drupes 1-3, on globose, 7-8 mm in diam. gynophore, subglobose, 2.6-3 cm across, brown, orange or yellow, tomentellous; pericarp woody when dry, 1 mm thick; endocarp bony, 2.2-2.5 cm in diam., with persistent calyx. Seeds whitish, subglobose with divaricate, much folded and divided cotyledons; hollow within, peltate, enclosing condyle; endosperm present. Embryo having very thin divaricate cotyledons with an irregular margin and a small superior radicle. Flowering season is from November-December Micromorphological studies on the leaf revealed that the dense tomentum is composed of two types of trichomes in the abaxial surface of *C. fenestratum* is a striking feature that distinguishes this species from other genera of the tribe *Coscinieae*.

Distribution

Geographical: India, Sri Lanka, Thailand, Cambodia, Vietnam, Thailand, Peninsular Malaysia, Sumatra, Bangka, western Java and Borneo.

Regional: States of Kerala, Karnataka and Tamil Nadu. In Kerala, it occurs in Thiruvananthapuram, Thrissur, Malappuram, Palakkad, Kollam, Idukki, Wayanad, Kannur and Kozhikode districts.

Climate and soil

This plant is an endemic species of Western Ghats, wherein it prefers rain forest climate and high humidity throughout with a temperature ranging from 15-25^{0}C. Plant occurs in partial shade in humid location in association with many shrubs and large trees. The crop requires acidic soil (pH of 5.2) with very high organic matter (2.3%) content.

Propagation material

The planting material is seed. The plant can be regenerated from stumps of old plants and also through used seeds, but the rate of regeneration is found to be extremely low. Seeds have a dormancy period of 6 months. The fresh stem cuttings of pencil size with about 15 cm length are suitable for vegetative propagation

Nursery technique

Raising propagules

The plant is propagated through seeds. Mature fruits occur in bunches from May to October. Fruit pulp could be removed by rubbing against rough surface. Black seeds may be soaked in 3000 ppm GA for 20 hours to hasten germination up to 90%. Seed should be immediately put for germination since they lose viability within two weeks. Seeds germinate very slowly, germination start after 4 months and may take even 8-10 months. Germination percentage of fresh seed is approx. 60%.

The plants from nursery bed are carefully shifted to polybags filled with mixture of sand, soil and FYM @ 2 kg/polybag. Plants are watered on alternate days and kept in shade in mist chamber. They are maintained for nearly 6-9 months and then shifted to main field after it has attained 60 cm in height.

Fruits and seeds of *Coscinium fenestratum* (Gaertn.) Colebr

Land preparation and manure application

Plant is raised over FYM. In all 20 kg FYM is applied in a pit and seedling is planted over the manure and nurtured well by watering regularly. The plant is given a bamboo pole to support for better plant growth.

Transplanting and optimum spacing

Well-grown plants are planted in main field along with soil attached to the roots in poly bags. Plant to plant spacing is 2m X 2m. Care is taken to put plant underneath a shrub or tree having 50-60 percent shade for quick establishment of crop.

Irrigation practices

Plant needs frequent irrigation and daily irrigation is required during summer months.

Crop maturity and harvesting

Crop attains maturity in about 15 years. The economic part is root. It is expected to give about 3-4 kg of fresh root/plant depending on the growth of plant and its environment.

Post-harvest management

The root is dried in shade and stored in air-tight gunny bags.

Chemical constituents

The major alkaloids are yellow crystalline berberine, protoberberine and jatrorrhizine. Many other alkaloids, mainly of the protoberberine type, isolated from stem and root are magnoflorine, berberrubine, thalifendine, palmitine and oxyberberine. The stem and root also contain ceryl-alcohol, saponin, hentriacontane, sitosterol, palmitic acid, oleic acid and sitosterol glucoside. Stem contains protoberberine alkaloids and jatrorrhizine.

Yield

The plant yields about 3-5 kg fresh root and reduced to half on drying. Thus, about 4 t dry roots can be expected from 2500 plants/ha.

Therapeutic uses

Though roots and stems of this plant are having medicinal properties, mostly dried woody stem of this plant is sold in the crude drug market. The dried woody stems of the plant constitute the drug. In action, root and stem of this plant is anti-inflammatory, antiseptic, antipyretic, antiperiodic, tonic and stomachic. The stem is used to treat tastelessness, bleeding piles, cough, wounds, ulcers, skin diseases, abdominal disorders, jaundice, liver disorders, intrinsic haemorrhage, diabetes, snake bite, fever and general debility. It is also used in cases of suppression of lochia, continued and intermittent fevers and certain forms of dyspepsia. The woody stem contains an alkaloid Berberine. Due to the antiseptic properties, it is used to dress wounds and ulcer. The stems yield a yellow dye, which is used singly or in combination with turmeric.

Substitutes and adulterants

It is stated that it may be used in place of Cinchona, Gentian, Columba and Berberis. Further, it is considered a fairly good substitute for Ceylon Columba [*Jateorhiza palmata* (Lam.) Miers], which is employed in veterinary medicines. Hence, this plant is referred to as false Columba. The root, root bark and stem of *Berberis aristata* DC (Berberidaceae) and the root, stem and stem bark of *Coscinium fenestratum* (Gaertn.) Colebr. (Menispermaceae) are used in Indian systems of medicine. However, the former drug plant is considered as genuine and the latter as substitute.

11

Kalimusali
Curculigo orchioides Gaertn

Hypoxidaceae

Ayurvedic name	:	Talamulika
Unani name	:	Musli Siyah
Hindi name	:	Kalimusali
Trade name	:	Kalimusali
Parts used	:	Tuberous roots and rhizomes

Field view of *Curculigo orchioides*

Morphological characteristics

Kalimusli roots are 5–22 centimetres long, cylindrical, tuberous, and 0.5–0.8 centimetres thick. Brownish on the outside, with noticeable transverse creases that are loosely spaced. Externally, lateral roots are robust, fibrous, dull white in colour, and spongy, with a length of 5 cm or more. When freshly cut, tuberous rootstock has a starch-white colour within and is mucilaginous. Simple, sessile leaves are clustered on a short stem with sheathing leaf bases that taper into a little petiole, giving the plant an almost radical aspect. They are 15–45 cm long and 1.2–2.5 cm wide, linear or linear–lanceolate, membranous, glabrous or sparsely soft haired, and are linear or linear–lanceolate, membranous, glabrous or sparsely soft haired. When the leaf tip comes into contact with the earth, it develops roots and creates adventitious buds.

Floral characteristics

Flowers are epigynous, bright yellow, bisexual or unisexual, with a lanceolate and membranous bract. The perianth is found at the top of a lengthy, sterile ovarian extension that extends above the ground. With six equal lobes measuring 1.5 cm and sparsely hairy along nerves on the interior, the perianth is gamotepallus. The ovary is tricarpellary, syncarpous, and trilocular, with a long, slender beak (stipe). Flowering and fruiting occur largely from October to January, with intermittent flowering and fruiting occurring all year.

Distribution

The species can be found in subtropical Himalayan shade zones, Western Ghats from Konkan southwards, West Bengal plains, Central India, Chhota Nagpur, and other tropical zones. It is a tropical plant that can be found in almost every region of India, from near sea level to 400 metres above sea level, especially in rock fissures and laterite soil. It is a shady plant that thrives in wet climates.

Propagation material

In February–March, 1.5–2 cm long tuber segments containing the apical bud are extracted and used for propagation.

Nursery technique

Raising propagules

No stock is raised in the nursery. At the commencement of the south-west monsoon, which passes over South India in May–June, tuber segments measuring 1.5 cm x 2 cm are harvested from mother plants and sown immediately in the main field. The tuber segments should be planted at a distance of 10 cm x 10 cm. In humid tropical places like Kerala, 70–80 percent sprouting is achieved after two months of sowing.

Propagule rate and pretreatment

The propagule rate is 600–750 kg of root segments per hectare. The tuber segments require no pretreatment before sowing.

Land preparation and fertilizer application

The crop thrives in damp, humus-rich environments. With the arrival of monsoon, the field is well ploughed. Before planting, organic manure is mixed with soil , and elevated beds are created to prevent waterlogging. At the period of land preparation, FYM (farmyard manure) is applied at a rate of 20 tonnes per hectare. Alternatively, 15 tonnes/hectare of FYM can be applied during field preparation, and 25:15:10 kg/hectare of NPK (nitrogen, phosphorus, potassium) can be used as top dressing during October–November. Instead of FYM, well-decomposed poultry manure at a rate of 2.7 tonnes per hectare, mixed thoroughly with the soil during field preparation, yields better.

Planting and optimum spacing

The tuber segments are planted in rows in the field. When tubers are sown in humid tropical places like Kerala, they germinate/sprout with a success rate of

70%–80% after two months. For a pure crop, an ideal spacing of 10 m × 10 cm or 10 cm × 15 cm, gives an optimum crop stand of 0.6–0.65 million which is desirable, while intercropping with a coconut yields a crop stand of roughly 0.2 million with a spacing of 20 cm × 25 cm.

Intercropping system

In the shade of irrigated coconut orchards, the crop thrives well. Artificial shade must be provided using shade nets of 25% density if it is to be cultivated as a pure crop.

Interculture and maintenance practices

Crop management does not necessitate the use of additional manure. Weeding is mainly done by hand. To keep the crop weed-free, weeding is done twice, two and four months after planting. Except for frequent weeding and watering during dry months, no extra upkeep is necessary.

Irrigation practices

During the monsoon season, the crop is grown in rainfed areas. It has to be irrigated with 5 cm floods every two weeks after the monsoon has passed.

Disease and pest control

Seedling rot can be controlled by spraying and saturating the soil with 1 percent bordeaux mixture during the rainy season. There is also evidence of black rot disease, which can be controlled by spraying 0.05 percent tridemorph. Because rhizomes are regularly consumed by rodents, standard pest control approaches can be employed to keep them at bay.

Crop maturity and harvesting

One month after planting, the plant begins to flower, with the most flowers appearing in the second and third months. Flowering can occur at any time of the year. Fruits and seeds, on the other hand, are not used in the production of pharmaceuticals. Roots after seven to eight months can be harvested by digging.

Cut rhizome pieces of *Curculigo orchioides*

Post-harvest management

Tubers are stripped of their remaining shoots and rootlets. The tubers are cleaned of dirt particles, then dried thoroughly in the shade before being kept in gunny bags.

Chemical constituents

C*urculigo* roots contain acetone extractives (1.5%–1.8%), ash (3.3%–3.9%), and curculigosides (0.2%). The rhizome is reported to contain glucose, mannose, xylose, glucuronic acid, glycoside, polysaccharides (hemicellulose and other polysaccharides), starch, resin, tannin, mucilage, fat and calcium oxalate

Yield

In well established plantations, it is possible to achieve a dried tuber yield of 1000–1700 kg/hectare.

Therapeutic uses

The rhizome is mucilaginous and sweet, cooling having properties like bitter, emollient, diuretic, aphrodisiac, depurative, alternative, appetiser, carminative, viriligenic, antipyretic and tonic. Bhavamisra described it as a rejuvenating and aphrodisiac drug. It cures morbid vata and pitta, improves complexion and is useful in general debility, deafness, cough, asthma, piles, skin diseases, impotence, jaundice, urinary disorders, leucorrhoea and menorrhagia.

Some of the commercial formulations containing *Curculigo orchioides* are kama sutra capsule (ALMA HEALTH CARE), vigorous capsule (TAMPCOL), strong-nite capsule (MEDIMIX), meno-peace capsule (D'ARCY NATURAL), potency plus (CHINESE), braincare 2000 (CHINESE), vaipani kamon (VAIPANI) and sharmiotone syrup (SHARMILA) which are claimed to be rejuvenative, energizer or aphrodisiac.

12

Black Zedoary
Curcuma caesia Roxb

Zingiberaceae

Plants of *Curcuma caesia*

Ayurvedic name	:	Narkachur
Unani name	:	Siyah haldi, Kali haldi
Hindi name	:	Kali haldi, Narkachur
Trade name	:	Black zedoary, Kali haldi
Part used	:	Rhizome

Some species of turmeric have been documented from China, Australia, and the South Pacific, despite the fact that turmeric has been farmed all throughout the tropics. Turmeric along with the 110 species is found growing in Tropical Asia, with India, Myanmar, and Thailand having the most diversity. Black turmeric (*Curcuma caesia*), along with *Curcuma aromatica*, *Curcuma amada*, *Curcuma aeruginosa*, *Curcuma longa*, and *Curcuma zanthorrizha*, is one of the economically important species native to India. Curcuma caesia (black turmeric/ black zedoary) is a critically endangered perennial herb in the Zingiberaceae family native to India. Traditional healers employ the *Curcuma* genus to treat a variety of diseases in ethno medical traditions, but *Curcuma caesia* Roxb. is a relatively unknown and unexplored medication. In the Indian subcontinent, black turmeric is frequently used as a spice, food preservative, and colouring ingredient. Black turmeric is the most recent spice to be chemically studied, with a few papers describing 100 or more secondary metabolites from this plant, and it will continue to be a source of new bioactive chemicals.

Distribution

Black turmeric is a native of Northeastern and Central India, where it has long been used in traditional rituals and medicinal treatments. Within an altitudinal range of 200–1000 m, the species is found in moist deciduous woods, primarily in Bengal, North East, and Central India. It develops as a woodland ground cover in subtropical to temperate climates.

Morphological characteristics

Flower and rhizome of *Curcuma caesia*

Rhizome: The rhizome is tuberous and has a camphoraceous pleasant odour. It is 2-6 cm in diameter with a varying form and size. It is sessile, with adventitious roots, root scars, and warts, and it's flattened laterally. Because of the circular wrinkles on the surface, the nodal and inter nodal zones are prominent. The rhizome surface is dark brown, bluish black, or buff in colour, with circular groupings of scaly leaf remains giving the appearance of growth rings. The branching is sympodral in nature.

Root: All over the surface of the rhizome are yellow brown long fibrous and tapered adventitious roots.

Leaves: The leaves are found in groups of 10-20. The leaves are glabrous and large oblong lanceolate. The lamina reveals strong farraginous purple coloured clouds in the centre section. The petiole is ivory in colour, and when it is unsheathed, it encircles the other petioles, making a pseudo axis. The variation is parallel, which is a common feature of monocots.

Floral characteristics

The inflorescence is a 15-20 cm tall thick spike that appears well before the leaf opens; the bracts are green, and the coma bracts are deep red; as it ages, it turns crimson. With a pale yellow and reddish border, the flower is smaller than the bracts. Corolla is long, tubular with pale yellow lip - 3 lobed semi-elliptic; Calyx is 10-15 mm long, obtuse, 3 toothed; The flowers bloom in June and July, and the fruits ripen in September and October.

Climate and soil

Curcuma Caesia thrives in damp deciduous areas. The crop is best planted in the spring or summer. It grows best in a warm, humid environment. It is a partly shade-loving plant, yet it thrives in full sun when grown in open condition. Soil that is rich in organic matter, wet, and well-drained is ideal. *Curcuma caesia* thrives in sandy loam soils with a pH of 4.5–6.5. It may be grown in a variety of soils, including light black, ashy loam, and red clay loam. It does, however, prefer a well-drained sandy or clay loam.

Land preparation

To achieve a good tilth, the land should be ploughed and harrowed. To avoid water logging during rainstorms, the field must be properly levelled and drainage measures should be installed. About 20 t of FYM or compost per hectare should be thoroughly mixed into the soil. The area is separated into 1 m wide beds, and rhizomes of suitable length are planted in shallow holes on the tops of the ridges. The ridges should be spaced 45-60 cm apart, and the plants should be spaced 15-30 cm apart. During field preparation, vermicompost, neem cake, trichoderma, and gypsum are combined at a rate of 10 tonnes per acre. If necessary, lime at a rate of 2 tonnes per hectare can be applied to soils to decrease excessive acidity at least one month prior to planting.

Nursery technique

Raising propagules

Rhizomes are used to propagate the crop vegetatively. Rhizome pieces are planted directly in the field, with no nursery stock in most cases.

Propagule rate

For planting at 30 cm x 30 cm spacing, approximately 1000 kg of rhizomes will be required per acre.

The rhizomes are cut into 4-5 cm long sections and planted whole or split into two portions, each with at least one sound bud. Depending on the weather, the crop is planted from April through July. In 15–20 days, the rhizomes will sprout.

Manures and fertilizer application

It needs heavy manuring. Usually, 40-50 tonnes of well rotten farmyard manure is applied at the time of preparing the land. For quality production, chemical fertilizers like 30 kg each of N, and P_2O_5 and 60 kg of K_2O per ha may also be applied in split doses

Intercropping system

Black turmeric is grown as a single crop. But, it may be intercropped beneath widely interrupted canopy trees, like Amla, Mango etc.

Intercultural operation

Mulching: Mulching the bed with green leaves is an important practice beneficial to this crop. Two or three mulching are given, the first immediately after planting and the second and the third at intervals of 40-50 days.

Irrigation: The crop is often cultivated in rain-fed conditions in Assam and Kerala states with substantial rainfall. Other regions will have to be irrigated on a frequent basis to maintain a constant humidity level. The finest irrigation method is sprinkler irrigation.

Weeding: Manual weeding is a common practice for controlling weeds. It should be done at regular intervals for the good growth of plants.

Disease and pest control

Occasionally, leaf spot (*Tephrina* sp., *Coletotrichum* sp.) and leaf blotch (*Corticium* sp.) could be seen affecting the crop. Spraying liquid organic booster at weekly intervals can help to control them.

Crop maturity and harvesting

It takes around nine months for the crop to mature. In mid-January, the harvest is completed. Before digging the rhizomes, the earth is wet with irrigation to prevent injury to the rhizomes. Rhizomes should be dug with extreme care and should be taken from the soil and cleaned after digging. The rhizomes are thoroughly cleansed to remove any soil that has adhered to them. The crop may be spoiled if the rhizomes are damaged. The rhizomes should be kept in the shade to dry after washing.

Chemical composition

Secondary metabolites from the black turmeric are predominantly camphor, terpenes, lactones, alkaloids, and phenols classification. The research on the volatile rhizomes oil of *Curcuma caesia* resulted in the identification of 30 components, representing 97% of the oil, with camphor (28%), ar-turmerone (12%), (*Z*)-ocimene (8%), ar-curcumene (7%), 1,8-cineole (5%), elemene (5%), borneol (4%), bornyl acetate (3%) and curcumene (3%) as the major constituents. Other research demonstrated diphenyl alkanoids, phenylpropene derivatives, terpenoids, flavonoids, steroids, and alkaloids as major constituents.

Post-harvest management

Peeled, half cut or sliced rhizomes should be kept in oven at 55 °C or under well-ventilated shade for drying. These dried rhizomes should be stored in suitable damp-proof containers.

Yield

Estimated yield of fresh rhizomes is 19-21 tonnes per acre while dry rhizome yield is about 3.5 to 5 tonnes per acre.

Value added products: *Curcuma caesia* Powder

Therapeutic uses

Black turmeric has been an essential component in general healthcare, particularly in the rural areas. It has been used by many tribal communities worldwide from centuries as spice, medicine and in spiritual practices. With advancement in technology, the species is gaining importance as a potential source of new drugs. *C.caesia* is a wonder herb and contains the highest content of curcumin and it is a chemical substance with many curative properties. It is used for treatment of menstrual disorders, piles, impotency and epilepsy. Externally, this plant has been used in the treatment of wounds, white patches on the skin and leprosy sores. It is also capable of enhancing fertility levels. It is also used for the treatment of enlargement of the spleen and certain types of tuberculosis. The rhizome as well as the leaves of the plant is used in medical formulations. The rhizomes are used as stimulants, anti-diarrheal, diuretic, anti-emetic, wound cleanser and in treating various skin disorders in India, Indonesia, Thailand and Malaysia. The paste of black turmeric rhizome is applied on wounds and in rheumatic pain in Manipur. In Arunachal Pradesh the fresh decoction of rhizome is used as anti-diarrheic while the fresh paste of rhizome is applied over snake and scorpion bite. These rhizomes are traditionally used in treatment of stomachache, typhoid and wounds in tongue by Monpa community of Dirang (West Kameng) in Arunachal Pradesh. Kali haldi is recognized as a medical herb with strong antibiotic properties. It is used as remedy for wound, bruises and sprain, skin problems, digestive aid, liver protection. It helps in reducing cholesterol. It is believed to play a role in preventing and curing cancer in Chinese medicine. It contains aromatic volatile oils that help to remove excessive lipids from the blood, reduce aggregation of platelets and reduce inflammation.

13

White turmeric
Curcuma zedoaria (Christ.) Rosc

Zingiberaceae.

Plant of *Curcuma zedoaria*

Hindi name	:	Kakhur
Trade name	:	Kasturimanjal
English name	:	Round zedoary
Parts used	:	Rhizomes and leaves

Curcuma zedoaria (zedoary or white turmeric) is a perennial herb and member of the genus *Curcuma*, family Zingiberaceae. The plant is native to South Asia and Southeast Asia but is now naturalized in other places including the US state of Florida. Zedoary was one of the ancient food plants of the Austronesian peoples. They were spread during prehistoric times to the Pacific Islands and Madagascar during the Austronesian expansion. Its use as a spice in the West today is extremely rare, having been replaced by ginger, and to a lesser extent, yellow turmeric. *Curcuma Zedoaria* (Berg.) Rosc.syn. *C. zerumbet* Roxb; *Amomum zedoaria* Christm., the Kasthurimanjal of commerce, is also known as Round zedoary in English, Kachura in Sanskrit and Kakhur in Hindi.

Morphological characteristics

Zedoary is a rhizomatous annual or biennial herb with a long, scented stem. It stands at 1.8 metres tall. On the inside, the rhizomes are large and pale yellow or yellowish. The roots are fragrant, and the tubers are ellipsoid in shape. With the exception of a bitter aftertaste, the edible root of the zedoary has a white interior and a mango-like scent; however, its flavour is more similar to ginger. Although there is no obvious aerial stem, the shoot has a pseudo stem made up of four to six long and densely overlapping sheathing leaf bases. The leaves are enormous, up to 1 m long, rectangular and strongly veined, and frequently violet in colour in the centre.

Floral characteristics

Flowers of *Curcuma zedoaria*

The rhizome give rise to an inflorescence, which is a spathe. Bracts are brilliant reddish-green and the flowers are whitish or pastel yellow. The funnel-shaped corolla tube is reddish. The calyx is pale and serrated in an obtuse manner. Flowering takes place in May–June, although fruiting is uncommon. Fruit is a capsule that is ovoid in shape.

Distribution

The species can be found in the forests of the eastern Himalayas, Bengal, and Kerala since it is semi-domesticated. It is grown in climate-friendly regions such as Kerala, Karnataka, Tamil Nadu, and other South Indian states.

Climate and Soil

Rhizomes of *Curcuma zedoaria*

The crop is a tropical, subtropical, and temperate semi-domesticated crop that thrives in moist climates. It thrives on loam and sandy loam soil with an average yearly rainfall of 1100 mm. It grows well in both open and shady areas.

Propagation material

The ideal material for propagation is rhizome, which can be obtained throughout the winter months (November–December). This crop does not necessitate seeding.

Raising propagules

The crop is not grown in the nursery. Rhizome fragments are directly planted in the field.

Propagule rate and pretreatment

Planting at a spacing of 40 cm x 20 cm requires 10–12 quintals of rhizome propagules per hectare. Before sowing, no special treatment is necessary.

Land preparation and fertilizer application

One disc harrow ploughing and two to three desi plough ploughings are required for soil conditioning. Before planting, the soil is carefully mixed with 150 quintals of FYM (farmyard manure) and 100:80:60 kg of NPK (nitrogen, phosphorus, potassium).

Planting

The crop can be planted anytime between April and June, depending on soil moisture availability. In the field, rhizome pieces are planted 20 cm apart in rows with a 40 cm row-to-row spacing. If the soil is damp, rhizomes grow after 10–12 days; otherwise, they remain dormant in the soil until the first rain. Planting requires around 125 000 propagules per hectare at a spacing of 40 cm × 20 cm. The output could be boosted if the spacing was a little broader.

Intercropping

This crop can be grown in either a solo or mixed cropping plan under partial shade from trees with a thin canopy.

Interculture and maintenance practices

At 30, 60, and 90 days following planting, the crop requires two to three weedings. After then, the plants have a weed-suppressing function. There is no need to apply any more organic or inorganic fertiliser.

Irrigation practices

The frequency of irrigation is determined by the soil and meteorological conditions. During the summer, the crop requires three to four light irrigations per month, while, during the winter, the crop needs two to three irrigations.

Disease and pest control

During the experimental trials, no evident symptoms of any disease, physiological ailment, or other condition were noticed or reported on *Curcuma zedoaria.*

Harvest management

In six to eight months, the entire crop matures. Rhizome harvesting takes place between November and December. These rhizomes are cleaned thoroughly with water to remove dirt particles, then sliced into slices and dried in the sun to eliminate surface water before being dried in the shade again. The finest months to harvest the leaves are October and November.

Chemical composition

Curcumol, curcolone, procureumenol, isocurcumenol, furadiene and its iso-derivative, curcumadiol, dehydrocurdione, and zederone are sesquiterpenes found in rhizomes. Essential oil comprising – pinene, camphor, and sesquiterphenes is obtained by steam distilling the dried rhizomes (more than 50 percent).

Post-harvest management

For commercial uses, sliced and dried rhizomes are kept in a cool spot. Rhizomes are buried in dirt pits or sand heaps until March to get planting material.

Yield

On a fresh weight basis, fresh rhizome yields range from 285 to 315 quintals/hectare, while leaf yields range from 70 to 80 quintals/hectare. The production of essential oil from leaves is 7–8 kg/hectare (on a 0.09 percent fresh weight basis), while the yield of essential oil from rhizomes is 80–90 kg/ha (at 0.28 percent of fresh weight).

Uses

The edible rhizome of zedoary has a white interior and a mango-like aroma. Its flavour, on the other hand, is more like to ginger, with a bitter aftertaste. In Indonesia, it is ground to a powder and added to make white curry pastes, whereas in India, it tends to be used fresh or in pickling. In Thai cuisine it is used raw and cut in thin strips in certain Thai salads. It can also be served cut into thin slices together with other herbs and vegetables with certain types of *nam phrik* (Thai chilli pastes). It is also used in India to flavour dals and chutneys. The plant is used traditionally to treat inflammation, pain, and a variety of skin ailments including wounds, as well as menstrual irregularities and ulcers. The essential oil produced from the dried rhizomes of *Curcuma zedoaria* is used in perfumery and soap fabrication, as well as an ingredient in bitter tonics.

Therapeutic uses

Rhizomes have fragrant, cardiotonic, carminative, diuretic, and antispasmodic properties. Jaundice, cough, hiccups, and respiratory problems are all treated with them. Aromatherapy and perfumery use essential oils collected from leaves.

14

Himalayan Marsh Orchid *Dactylorhiza hatagirea* (D.Don) Soo

Orchidaceae

Ayurvedic name	:	Salampunja
Unani name	:	Buzidan, Salab Misri
Hindi name	:	Salampanja
English name	:	Himalayan Marsh Orchid
Trade name	:	Salampanja
Parts used	:	Tubers, roots

Inflorescence of *Dactylorhiza hatagirea*

Morphological characteristics

Plant is terrestrial glabrous herb attends height of 20-25 cm. Tubers slightly flatten ed, palmately divided into 3-5, finger like lobes. Flower stalk upto 90 cm Long, erect, hollow, leafy throughout or the lower portion bear few sheathing scales. Leaves are 4-6, cauline, leaf blade oblong to linear lanceolate, 8.0-15.0cm x 1.5-3.0 cm in size, base sheathing, apex obtuse or acuminate.

Floral characteristics

Inflorescence is raceme, 5.0-15.0 cm long, crowded with many flowers. Flowers are purple and the bracts green, narrowly lance shaped, lower bract longer than the flowers, upper slightly shorter. Flowers are about 1.8 cm long, including the curved spur. Sepals and petals are nearly equal. Three of them form a hood and the two sides spread outwards. The lip is rounded and shallowly 3-lobed, spotted dark purple. Spur straight, cylindrical, nearly as long as the ovary; column very short; anther adnate to its face, cells diverging; pollinia 2, caudicles attached to 2 small globose, viscid glands enclosed in a minute pouch overhanging the broad, 2-lobed stigma.

Distribution

This attractive terrestrial orchid grows in damp pastures of Himalayas from Kashmir to Nepal at the altitude of 2500-5000 m MSL.

Climate and soil

Plants grow in the alpine climate of Central and Western Himalayas where average summer temperature ranges between 15 -18 °C and winters are very harsh generally below the freezing point. Plants grow best in moist meadow soils. These soils are usually dark grey, granular, sandy loam, micaceous sandy soils at greater depth.

Propagation material

The materials for propagation are seeds and tuber cuttings. Plants can be grown by splitting the sprouting tubers.

Collected seeds did not show germination at controlled temperature and moisture (temp. 10-15^{0}C; humidity 90+5%) conditions. It probably requires thawing under the snow for about 4-5 months and mycorrhizal association for germination. It is reported to be an inherently slow growing and poorly regenerating species because of pollinator specificity.

Nursery technique

Raising Propagules

Vegetative propagation through tuber cutting is quite successful in this plant. Small slices of tuber of 4.0 mm size, with meristematic tissues are reported to develop plantlets when transplanted at 5.0cm -7.0cm depth at a spacing of 15cm X 15cm. Plant raised from tuber cutting develops tubers and roots after one year and it becomes about 4. 0-8.0 cm tall.

Acclimatization of collected saplings in the glass house

Propagule rate and pretreatment

Multiplication through seeds is very low and no fruitful results have been obtained so far. In natural habitat, seed germination was found to be between 20 to 30%. Seed sowing in soil collected from its natural habitat at lower altitudes gives better results. Collection of immature and fresh seeds from nature enhanced germination percentage. Approximately 1,11,150 tubers or tuber segments are required for one hectare of land and these are transplanted at spacing of 30 cm apart.

Land preparation and manure application

Land should be ploughed and cross ploughed to have a fine tilth. The field should be well prepared and made free from weeds before transplanting. Livestock manure and forest litter treatment increase survival, growth and yield. Approximately 5t manure is required for one hectare of land. Manure requirement increases at lower altitudes and at 1800-2000 m approximately 10.0 - 15.0 t/ha manure is required depending upon soil fertility.

Transplanting and optimum spacing

Small slices of tuber of 4.0 mm size with meristematic tissues are reported to develop plantlets when transplanted at 5.0 -7.0 cm depth at spacing of 15cm X 15cm.

Intercropping system

Intercropping with A. *heterophyllum* Wall. ex Royle is found suitable because of similar edaphic conditions for good growth and yield. Intercropping with *Swertia chirayita* Roxb. ex Fleming is also recommenced for subsidiary income from the cultivation of medicinal plants.

Intercultural and maintenance practices

Frequent weeding at every seven to ten days especially during the rainy season favour optimal growth.

Irrigation practices

At the early stage, 80-90% humidity is required for plant development of rooting and leaf initiation. Irrigation at every twelve hours is needed, especially at lower altitude during this stage.

Chemical constituents

Tubers of *D. hatagirea* are rich in starch, mucilage, sugar, phosphate, chloride and glucoside-loroglossin. *D. hatagirea* is also used to treat bone fractures in Uttarakhand. *D. hatagirea* tubers are known to produce a high-quality 'Salep,' which is widely utilised as a nervine tonic in local medicine due to its astringent and aphrodisiac characteristics.

Crop maturity and harvesting

Tubers are usually harvested after five years to fetch a good yield. Sometimes it is harvested after two or three years and the tubers are used as food, nervine tonic and aphrodisiac.

Tubers of *Dactylorhiza hatagirea*

Therapeutic uses

The crop is commonly used to treat dysentery, diarrhoea, chronic fever, cough, stomachache, wounds, cuts, burns, fractures, and overall weakness, especially in debilitated women after childbirth, as well as to boost regenerative fluids The tuber juice is used as a tonic and as a pyorrhea therapy. Root paste is applied to cuts and wounds as a poultice, and the extract is used to cure stomach issues. The tubers are eaten, used as an aphrodisiac, and used as a nervine tonic.

15

Rattalu, *Dioscorea bulbifera* L

Dioscoreaceae

Leaf, Tuber and the Whole plant of *D. bulbifera*

Ayurvedic name : Varahikhand, Varahi

Unani name : Zaminkand, Ratalu

Hindi name : Ratalu, Genthi

Trade name : Rattalu

Parts used : Corms and tubers

The genus Dioscorea comprises of approximately 600 species and sustains livelihood around the globe. Air yam, aerial yam, bitter yam, cheeky yam, potato yam, and parsnip yam are all names for *Dioscorea bulbifera* L., a traditional medicinal plant native to Asia, Northern Australia, America, and tropical Africa. It's been used in traditional medicine for a long time and has a lot of potential.

Morphological characteristics

It is a perennial herbaceous climber that can reach a height of 10-20 metres and has multiple fibrous roots. The leaves are alternating, broadly cordate, and bright green in colour, with rounded basal lobes arising from one of them. Aerial tubers (bulbils) are circular, smooth-surfaced tubers that grow freely in leaf axils. Annual stems emerge from a tuberous, woody rootstock. These stems either creep along the ground or climb up the surrounding vegetation. Dioscoreaceae subfamilies are distinguished by the presence of bisexual and unisexual blooms. Seeds are trilocular, dark brown, and somewhat winged (1.2-1.6 by 0.5 cm).

Floral characteristics

Flowers are grouped in spikes that are hanging, axillary, solitary, or fascicled. The seeds are winged at the base, while the capsules are oblong and winged. The fruit has three valves and the perianth is tubular. From September to November, the plant blooms and bears fruit.

Geographical distribution

Tropical Asia, Northern Australia, America, and Sub-Saharan Africa are all home to this plant species. The species can be found in tropical and subtropical environments up to 1000 metres above sea level. It is a shade-loving species that may also thrive in open places.

Climate and soil

D. bulbifera prefers to grow in areas with high temperatures (around 25-35°C), high humidity and high precipitation (>1000 mm annual rainfall), at low to middle elevations. This species grows best in a loamy soil that has good drainage and is preferably high in organic material. *D. bulbifera* is adapted to partially to fully shaded conditions, but it does not tolerate salty or frosty conditions and plant vigour is affected at temperatures below 20°C. It is more tolerant to low temperatures during dormancy. As with many species within the genus *Dioscorea*, *D. bulbifera* does not support its own weight to any great height & in order to capture sunlight, this species has evolved to climb by twining on other plants. Sandy loam soil is most suitable for its cultivation & better yield of corm.

Propagation material

Underground corms or aerial bulbils are both effective ways to propagate the plant. Plants that are propagated vegetatively grow more quickly than seedlings. Bulbils with a diameter of more than 4 cm sprout two days earlier than those with a diameter of less than 4 cm. The months of November and December are good for collecting bulbils for propagation. Bulbils are protected by being buried in the ground and used as propagules for the following season.

Nursery technique

Raising propagules

The preferable procedure is to plant the corms or bulbils directly in the field rather than transplanting nursery-raised seedlings. In the mid-hills, April–May is the best period to sow corms or bulbils. Seeds are separated from mature fruits that are collected in November and sowed in the spring or early summer to raise seedlings. Even under laboratory circumstances, germination takes 15–30 days and the germination per cent is less than 40%.

Propagule rate and pretreatment

About 80–100 kg well-developed bulbils or corms are required for plantation on 1 hectare of land at a spacing of 50 cm × 50 cm. No pretreatment is required before sowing the bulbils.

Land preparation and fertilizer application

The soil should be well pulverised and the planting sites should be properly tilled and weed-free. At the time of pre-planting tillage, a basal dose of 15–20 tonnes/hectare of FYM (farmyard manure) should be applied to the soil.

Planting and optimum spacing

Plant spacing of 50 cm x 50 cm is said to be ideal for good growth and productivity. This spacing results in a crop stand of 40, 000 plants per hectare, which is ideal. Plant spacing of 100 cm x 100 cm produces more bulbils per plant, but the average weight of bulbils as well as their number per unit area are substantially higher at 50 cm x 50 cm.

Intercropping system

The plant is preferred as a solo crop, but needs staking support or host of shrubs and trees.

Interculture and maintenance practices

When organic manure is applied at a rate of 15–20 tonnes per hectare at the time of planting, the average weight and production of bulbils increases. In the absence of sufficient FYM/organic manure, appropriate quantities of inorganic fertilisers can be used to compensate for the lack of FYM and increase output. To expose as much leaves to the sun as possible, the twiner requires staking support. A system of bamboo pillars and string, as well as wooden posts or trellis, may be used to support the plants.

Irrigation practices

Sprinklers should only be used for gentle irrigation to maintain humidity throughout the dry season. Flood irrigation can cause waterlogging, so it is best to avoid it.

Weed control

Manual weeding at monthly intervals during the peak period of weed growth (July–September) controls weeds to a great extent.

Disease and pest control

No serious disease or infestation has been reported in this crop.

Crop maturity and harvesting

Bulbils are ready to harvest at 180–200 days after they sprout. After two to three years, the mother corm can be harvested. During the winter, the crop falls dormant. As a result, the months of October–November are suitable for harvesting bulbils/underground corms. Mid-September to mid-October is the best time suited for subtropical and tropical areas. If the bulbils are not harvested when they are completely grown, they will be shed before the foliage develops.

Post-harvest management

The corms and bulbils can be stored in gunny or paper bags and the sprouting rate will not be affected. There could be a two-day delay in the sprouting of the bulbils.

Chemical constituents

Chief active ingredient of corm is diosgenin along with furanoid norditerpenes, sinodiosgenin, diosgenin, smilagenone, and epismilagenins.

Yield

The crop yields an average of 12 tonnes of fresh corms per hectare in the second year.

Therapeutic uses

Cough, epistaxis, goitre, hemoptysis, pharyngitis, skin infections, piles, throat infections, and dandruff removal are all addressed by the use of *D. bulbifera*. In Chinese traditional medicine, it is also used to cure a variety of diseases, including poison detoxification, blood clotting to stop bleeding, and clearing pathogenic heat and cancer. Cough, dysentery, piles, ulcers, diabetes, leprosy, and syphilis are all treated with *D. bulbifera* tubers, which are roasted and cooked like a vegetable. Tubers of *D. bulbifera* are boiled and eaten by locals in Uganada to treat HIV patients. Raw bulbil of *D. bulbifera* is used to treat ringworm in the Democratic Republic of Congo andis used to treat diarrhoea, struma, dysentery, and throat infections, tuberculosis in Indian traditional medicine.

The Bhoxa community in Dehradun, Uttarakhand, uses the tubers to cure diarrhoea and dysentery. To treat cough, roasted and mashed tubers are administered orally with salt. To treat dandruff, the twigs and delicate shoots are crushed and put to the hair. The leaves of are widely used to treat skin disorders in traditional medicine. Furthermore, the tubers are consumed uncooked to increase hunger.

16

Shankhapuspi
Evolvulus alsinoides Linn.

Convolvulaceae

Ayurvedic name	:	Vishnugandhi, Shankha-pushpi
Unani name	:	Sankhaholi
Hindi name	:	Phooli, Sharikha-pushpi
English name	:	English Speed-wheel
Trade name	:	Shankhapuspi
Parts used	:	Whole Plant

Evolvulus alsinoides - the plant

Morphological characteristics

It is an annual/perennial herb with prostrate spreading branches in all directions. The root system is deep. Leaflets are 2.5-5 cm long, elliptical, oblong and rounded at base. Leaves are densely clothed with white appressed and long spreading hairs. In nature, seedlings appear after 2-3 showers during July-August. Generally, the seedlings are abundant under the canopy of trees or shrubs. The young seedling grows straight and produces lateral branches during first week of August. The lateral branches spread in all directions. The flowering starts during first week of August. Fruit formation takes place in the last week of August, while flowering still continues. The dispersal of the seeds takes place from September to December. Plant dries during November-December.

Floral characteristics

Flowers are blue in colour and axillary in position. They are either solitary or in pairs on the long stalk. Styles are two and each is bifid. Fruit is globose with four-valved capsule containing dark brown to black smooth seeds. The flower colour varies from dark purple to whitish-purple.

Distribution

Plant is commonly found growing in open and grassy places, almost throughout India, ascending upto 1500 meter in the Himalayas. It is cultivated also in the sub-tropical parts of India, ascending upto 1800 meter elevation.

Climate and soil

The plant prefers shady and humid climatic conditions. Growth is slow and becomes restricted when the environmental conditions become unfavourable.

Propagation material

Seeds : Collected during October-November from natural habitats.

Nursery technique

Raising propagules

Mechanically scarified seeds are sown @ 2 seeds in each polybag at 0.5 cm depth during June-July to raise nursery.

Propagule rate and pretreatment

It has hard seed coat, for the removal of hard seed coat; the seeds are subjected to mechanical scarification which gives 80% germination. 190 gm seeds are required for planting one hectare area.

Land preparation and fertilizer application

The land should be prepared as for vegetable growing and made weed free and planked to a uniform level. For maximum yield, farm yard manure at the rate of 10 tonnes per hectare should be applied.

Transplanting and optimum spacing

The seedlings are transplanted during late July-August at 25X25 cm spacing under field conditions.

Inter cropping

Study revealed that inter cropping did not cause any adverse effects on the growth of *Commiphora wightii* and *Evolvulus alsinoides,* when both are grown together. Non-significant results were observed when this plant is intercropped with other crops/medicinal plants such as *Pennisetum glaucum* (Pearl millet), *Cyamopsis tetragonoloba* (Guar bean), *Vigna radiata* (Moong bean), etc.

Irrigation

Five days irrigation schedule was found to be suitable for obtaining maximum growth and biomass yield under dry field conditions. However, in case of rains, the irrigation schedule may be changed accordingly.

Weed control

Manual hand weeding and hoeing is required at an interval of 15-20 days during entire crop season.

Disease and pest control

No serious insects and pests are observed on the crop.

Crop maturity and harvesting

It is a rainy season crop and takes about four months to complete its life cycle. The seedlings transplanted in the month of July attain its full growth during September-October. This is the appropriate time for harvesting the plant.

Post-harvest management

Fresh herbage is perishable and should be stored in gunny bags for marketing after proper drying.

Chemical constituents

Plant contains an alkaloid evolvine; â-sitosterol, stearic, oleic and linoleic acid, betaine, pentatriacontane and triacotane.

Yield

As a pure crop, 18.6 quintal (fresh herbage) per hectare is obtained which is reduced to 1/3rd after drying. The dry herb is stored in cool and dry places.

Therapeutic uses

The whole plant is used in form of decoction in nervous debility and loss of memory. The plant is also useful as blood purifier and in bleeding piles. The fresh flowers with sugar are eaten as a brain tonic. The leaves are made into cigarettes and smoked in chronic bronchitis and asthma. It also improves complexion, voice and cures from intestinal worms. It promotes 'medha', the power of memory.

17

White Teak
Gmelina arborea Roxb

Verbenaceae

Ayurvedic name	:	Gambhari
Hindi name	:	Gamhar, Khamer, Gambhari, Sewan
English name	:	Candahar tree, White Teak
Trade name	:	Gamhar, Gambhari
Parts used	:	Root, leaves, bark and fruit

Morphological characteristics

It is a moderately deciduous tree with straight trunk and numerous spreading branches, which form large shady crown with whitish grey corky lenticellate bark, exfoliating in thin flakes. It has a clear bole of 6.0-9.0 meter, and a girth 1.5-2.5m. Branchlets and young parts are clothed with fine white mealy pubescence. Leaves are simple, opposite, broadly ovate, cordate, glandular, glabrous above when mature and fulvous-tomentose beneath. It attains its largest dimension in the mixed forests of moist region as in the eastern sub Himalayan track, Assam and elsewhere in south India.

Floral characteristics

Flowers are 2.5-5.0 cm in diameter, brownish yellow that ends in terminal panicle. Calyx is campanulate, pubescent outside and with 5 lobes. Corolla is showy, brownish yellow with short tube and oblique limbs. Stamens are 4 in number and didynamous. Ovary is 4 chambered with one ovule each; style slender ending in a bifid stigma. Fruits are drupe and obovoid in shape, 1.8-2.5 cm long, pericarp leathery, shining yellow outside when dry; endocarp bony, embedded in an aromatic sweetish pulp. Seeds are 1 or 2, hard and oblong ex-albuminous. The panicles of flowers appear from February to April.

Distribution

Seedlings of *Gmelina arborea*

Though it is a native of Asia, it has extensively been planted in tropical areas of Africa, Asia, Australia, America, West Indies, and on several islands in Pacific Ocean. In India, it is naturally distributed in Eastern SubHimalayan tract, Aravali Hills, Western Peninsula, Indo-Gangetic plains, Western Himalaya and Central India. It occurs naturally from 5° to 30°N longitude and between 50 to 1300 m elevations.

Climate and soil

It thrives well in shade in the temperature range of 30^0C – 47^0C, 60-100% humidity and the annual rainfall between 750 mm to 4500 mm. Its choice of site is wide, but it shows a preference for moist fertile valleys with sandy loam soil. It does not thrive in waterlogged conditions and remains stunted on dry, sandy or otherwise poor soil. It also does not thrive on heavy clay soils. *G. arborea* is a light demanding, moderately frost hardy species and comes up well with a temperature range of 30 to 48°C and mean annual rainfall of 760 to 4500 mm, grows well in soil pH ranging from 5.0 to 8.0 with preference for moist fertile alluvial with sandy loam soil.

Propagation material

Seed is used as the planting material here.

Nursery technique

It is a light demanding tree. It does not tolerate drought but it withstands in light frost condition. Seed formation occurs in May-June. Seeds are dried well before germination.

Raising propagules

Seeds are dibbled at spacing of 7.5 cm X 7.5 cm, 1 to 2.5 cm deep in unshaded nursery beds. The bed should be slightly raised to ensure good drainage. Seeds germinate in 20-25 days. During this period, weeding and watering is done according to the requirement.

Propagule rate and pretreatment

Dried seeds are soaked in water for 1-2 days to accelerate germination. Considerable amount of heat and moisture is needed to stimulate the germination and so the beds are covered with a layer of hay. The average number of seeds/ kg is 2000 - 2500. The germination percentage of seed recorded is 13-90%.

Land preparation and manure application

The land is prepared before planting and the unwanted herbs, shrubs and trees are removed. It is better to do deep ploughing to loose the land mass and allow to dry the unwanted weeds. Pits of 45 cm X 45 cm X 45 cm of size are dug during the month of May at a spacing of 4m X 4m. The pits are filled with well matured Farm Yard Manure (FYM), sand and soil in the ratio of 1:1:1 and allowed to cure before undertaking plantations in the month of June-July after onset of the rainy season. The choice of site is wide, but it shows a preference for moist fertile valleys with sandy loam soil.

Transplanting and optimum spacing

About 10-15 cm tall seedlings are transplanted in pits at the beginning of rainy season. The optimum spacing recommended between plant to plant is 4m X 4m.

Interculture and maintenance practices

Plants raised in the field require two weeding around the pits especially in rainy season. Plantation done in the black cotton soil require four weeding at monthly intervals. The unwanted weeds between the rows are removed by sword or sickle. Cultivator ploughing through tractor is beneficial to remove the weeds as well as loosening the soil especially in the black cotton soil. This operation not only conditions the soil but also avoids cracking associated with the black cotton soil during the summer months. It is advisable to run the cultivator before the start of active summer months to avoid cracking of soil to save water and damage of roots. Under natural conditions germination takes place in the rainy season soon after the fall of fruits. Heat and moisture stimulate germination. The hard coat of seeds takes time to rot and germination takes place only in the next rainy season. Crop can be raised through direct sowing and transplanting. Seeds are sown in lines at a distance of 3.0mX3.5m. Saplings are thinned in the third year. Dibbling and broadcasting on prepared sites also gives satisfactory results. Weeding hoeing, soil working and manuring is essential in the first year. Though the artificial regeneration mainly depends on seedling raised from seeds, cutting also strikes roots well. Growth is faster in case of vegetative propagation usually done through stumps and cuttings.

Irrigation practices

Weekly irrigation is required in summer season and irrigation at fortnightly interval is preferred in winter. Irrigation is required in the initial two years of the establishment of plants.

Weed control

One or two weeding in the month of July and September is enough for establishing the plantation.

Disease and pest control

The common nursery disease reported is sooty mould caused by *Corticium salmonicolor* which can be controlled by applying a suitable fungicide. The trees are often attacked and completely defoliated by the beetle (*Calopepla leayana*). It feeds on the leaf and also eats young buds and shoots. Defoliation is first noticed at the beginning of the rains and continues till October. No biological or mechanical control measures have been developed. Larvae of other several insects are known to bore in the wood and to defoliate, but the damage caused is usually not appreciable. Among fungi the one which are known to cause damage to this tree are *Poria rhizomerpha*, commonly a saprophyte but becomes pathogenic when G. *arborea* is raised in clayey soils which become water logged periodically, causing brown cuboidal rot in the roots resulting in die-back and death of affected plants. Trees are also liable to severe attack by Loranthus (*Loranthus scurrula*) as the tree has a thin bark, it readily becomes a victim to phanerogamic parasite.

Leaves of *Gmelina arborea*

Bark of *Gmelina arborea*

Crop maturity and harvesting

Tree grows fast and may be ready for harvesting of bark after 7 years. This plant is coppiced and traded. The roots are also harvested for medicinal purposes. The tree may stand up to 25 years. The medicinally important part of this species is stem bark which is extracted from 7-10 years old tree. Since, this is destructive harvesting method; it is suggested that from one tree partial debarking should be done by removing bark in patches of 15cm X 15cm with a distance of 60 cm. For getting roots, from the young plant it is desirable that the root should have good thick bark so as to get maximum active principle. Since, harvesting of roots and bark would be destructive, it is recommended to collect the bark from the clear felled crop as secondary product to avoid destruction of the plants growing in nature. Yellowish green fruits are collected from April-June from the ground duly rejecting the green and black ones. Fruits are heaped under or buried in a pit for 4-5 days and then washed to remove the pulp.

Post-harvest management

Bark that has been properly dried and has a moisture level of less than 10% can be stored in gunny bags in a well-ventilated area. Bark having moisture is susceptible to fungi infestation which turns it black in colour and becomes useless for medicinal use. Dried seeds can be stored in air tight container for one year and with this viability is decreased to a great extent.

Chemical constituents

The fruit contain butyric acid, tartaric acid and traces of resinous and saccharine substances. Arboreol is the major active constituents present in the plant. The roots contain thick yellow coloured oil. It also contains alkaloids, sugars, resins and some astringent compounds.

Therapeutic uses

Gmelina arborea is a common species in traditional medicine with usefulness to treat many disorders. It possesses a wide spectrum of pharmacological properties such as wound healing, cytotoxic, cardioprotective, antimicrobial, anti-inflammatory, antioxidant, antipyretic, analgesic, and antidiabetic potential. Root of *Gmelina arborea* is an ingredient of the **"Dasamula"**.It promotes digestive power and improves memory. Roots are useful in fever, dyspepsia, haemorrhoids, stomachalgia, heart diseases, nervous disorders, piles and burning sensation. Bark is used in fever and dyspepsia.

18

Gurmar
Gymnema sylvestre R. Br. ex Schult.

Asclepiadaceae

Ayurvedic name : Meshashringi

Unani name : Gurmar buti

Hindi name : Gurmar

Trade name : Gurmar

Part used : Leaf

Diabetes mellitus is a severe endocrine illness that affects almost 10% of the world's population and is a serious public health concern. In its most severe form, the disease affects major body systems, resulting in multiorgan problems. Oral hypoglycemic medicines such as sulphonylureas and biguanides are the most commonly used treatments, although their negative side effects are a significant drawback. Herbal medications are gaining popularity as a result of their superior outcomes and safety when compared to commercially available drugs, as well as their ability to address health problems more effectively.

Plants with anti-diabetic qualities are of particular interest to the ethnobotanical community since they are known to contain useful medicinal properties in various parts of the world, and a number of them have demonstrated variable degrees of hypoglycemic and antihyperglycemic action. Several plant species that are high in terpenoids, flavonoids, phenolics, coumarins, and other bioactive compounds have been found to lower blood sugar levels.

Gudmar (*Gymnema sylvestre* Retz) is a valuable medicinal woody climber from the Asclepiadaceae family. The medicinal climber is well-known for its anti-diabetic qualities. Its formulations have potent taste modifying properties, reducing sweet taste sensations in particular. *Gymnema sylvestre* is known in Hindi as gurmar, which means "sugar destroyer." The plant has become vulnerable as a result of increased demand and destructive harvesting. The genus is classified into 40 species, some of which like *G. sylvestre, G. montanum,* G. *yunnanense*, and *G. inodorum* have medicinal properties. The leaf extracts contain gymnemic acids, the major bioactive constituents that interact with taste receptors on the tongue to temporarily suppress the taste of sweetness.

Distribution

It is widely found growing in the tropics of Africa, Asia and Malaysia. In India, this plant is found growing in abundance in the forests of Karnataka, Tamil Nadu and Bihar. Due to its raising demand in South East Asian countries, the plant is becoming endangered particularly in Tamil Nadu.

Morphological characteristics

G. sylvestre is a slow growing, perennial, medicinal woody climber found in central and peninsular India. *G. sylvestre* is a large, more or less pubescent, woody climber. It is occasionally cultivated as medicinal plant. Leaves are opposite, usually elliptic or ovate (1.25–2.0 inch × 0.5–1.25 inch). Flowers are small, yellow, in umbellate cymes. Follicles are lanceolate, upto 3 inches in length.

Soil and climate

The plant grows in a variety of soil and agro-climatic conditions in tropical and sub-tropical regions up to 600 m elevation. Red sandy loam or medium deep black soil are reported to be ideal for this crop. The plant is sensitive to water logging and hence its cultivation on such soil should be avoided. Madhunashini prefers tropical and sub-tropical type of climate. It is found growing even in dry areas also. The areas with high or medium well distributed rainfall are suitable for its cultivation.

Varieties/types

Based on the leaf size. The climber can be classified into two types:

a) Small leaves type: Leaves are oval measuring 1.0-3.5 cm length and 1.5 – 2.5 cm very soft, found in dry regions.

b) Broad and pubescent type: These leaves are also oval measuring 3-6 cm in length and 3.5-5.0 cm in width. Leaves are dark green compared to small leaves type and also are pubescent.

Nursery raising

Mature seeds are collected between October-December and sown in polyboxes/ bags or small plots as nursery. The seedlings raised are transplanted in field during February-March. The plant grows well with the on-set of rainy season. The climber is given proper support for its better growth and development. It can also be planted in between trees as intercropping. The plant can also be propagated through cuttings and planted during rainy season.

Different experiments were carried out to standardize seed germination, vegetative propagation and cultivation practices of the species. The study revealed that the species can be cultivated successfully by seeds as well as by rooted cuttings at a spacing of 50 x 50 cm. It was observed that pretreatment of seeds in cold water for 24 h improves the seed germination. For vegetative propagation, hard wood cuttings of 10-15 mm diameter having three nodes were found most suitable. March and July was found the best time for vegetative propagation. Among hormonal treatments, dipping of cuttings in 500 ppm IBA solution for 30 min was suitable for maximum rooting (52.50%).

Land preparation and planting

Gudamar is an evergreen climber and the best season for planting is June-July. After the ploughing and leveling of the land, 45 cm^3 sized pits are made at a distance of 2.5 m between the rows and 1.75 m between plants (within the row). The pits are dug open 15 days earlier to planting, they are filled with green leaves and top soil and 2 kg of well rotten manure per pit is added. The pits are to be irrigated and left for one week, then the rooted cuttings are planted in the pits.

Irrigation

Irrigation is given immediately after planting and fertilizers are applied. Later on, irrigation once in 5-6 days is sufficient. During summer depending on the weather condition irrigation frequency needs to be increased.

Weeding and hoeing

Periodical weeding and hoeing is done during and after rainy season. On proper establishment of plants weeding is done on quarterly intervals.

Insect pest and diseases

Major insects: Thrips, mites and green fly

Major diseases: Powdery mildew and leaf spot

Management

1. Spraying of the vines with 2 ml of Rogor per liter of water can control thrips. Mites can be effectively controlled by any Acaricide. To control green fly spraying of 1 ml nuvacron mixed in one liter of water.
2. The diseases can be controlled by spraying 3 g of water soluble sulphur (or) 2 g of Mancozeb in one liter of water at an interval of 10-15 days.

Chemical composition

The plant is known to have gymnemic acids A, B, C and D which have the gymnemagenin and gymnestrogenins. Besides, leaves also contain nonacosane and hentria contane which were isolated from the hexane extracts of leaves. An alkaloid gymnamine, which is a trace constituent has been isolated and identified. Besides this, other plant constituents are flavones, anthraquinones, hentri-acontane, pentatriacontane, á and â-chlorophylls, phytin, resins, *d*-quercitol, tartaric acid, formic acid, butyric acid, lupeol, â-amyrin related glycosides and stigmasterol. The anti-diabetic property of the plant is attributed to the presence of mixture of triterpines and saponins in the leaves.

Harvesting and yield

The crop will be ready for harvest after two years of planting. Leaves are the economic part and the harvesting of the leaves is done twice in the year. First harvest is done when the plant started flowering, *i.e.* during June and second harvest was done in September-October. Leaves are harvested by hand plucking and the main stem is not cut during harvest of leaves unlike the prevailing practice. The harvested leaves are dried under shade by allowing sufficient air to circulate by spreading thinly on clear ground for about 7-8 days. Direct sunlight is avoided to maintain quality of the leaves. The crop is harvested only once in a year during flowering and on an average 5-6 kg dried leaves per plant can be obtained from a 4 years old plant yielding about 10,000 – 15,000 kgs of dried leaves per hectare. The crop can be cultivated for 10-15 years under good management.

Value added products: Gudmar Powder

Therapeutic uses

G. sylvestre is a native herb that belongs to the Asclepiadaceae "milk weed" family. A great number of bioactive compounds can be found in the plant. It has a long history as one of the most important botanicals used in Ayurvedic medicine to treat everything from diabetes to malaria to snakebites. Chigengteng or Australian Cowplant is another name for the herb, which is also known as Waldschlinge in German and periploca of the woods in English.

Furthermore, it has substantial potential in major health issues such as cardiovascular disorders, obesity, osteoporosis, rheumatism, cough, dyspnea, ulcers, eye pain, and asthma, besides being a popular drug for a variety of other ailments. The herb is used in a variety of food preparations to regulate obesity and blood cholesterol levels, as well as to maintain sugar homeostasis.

The herbal preparations of *G. sylvestre* are presently used in tea bags, health tablets and supplements, beverages, and confectioneries. It also has antibacterial, antihypercholesterolemic, hepatoprotective, and sweet-suppressing properties. It also is utilised as a feeding deterrent for the caterpillar *Prodenia eridania*, to prevent *Streptococcus* mutans-caused dental cavities, and in skin cosmetics. Colgate Diabetic Toothpaste is a specially formulated ayurvedic toothpaste that contains. It has properties that include astringent, stomachic, antiperiodic, diuretic, and tonic. The leaf powder stimulates heart and circulatory system and increases secretion of urine and activates uterus. The drug enters into the composition of preparations like Ayaskrti, Varunadi, Kashya, Varunadighrtam and Mahakalyanakaghrtam etc. Roots have been reported as a remedy for snakebite.

19

Puskarmool
Inula racemosa Hook. F

Asteraceae

Inula racemosa - field view

Ayurvedic name : Puskara
Hindi name : Puskarmool
Trade name : Puskarmool
Parts used : Root and rhizome

Inula racemosa is an Asian plant in the daisy family native to the temperate and alpine western Himalayas of Xinjiang, Afghanistan, Kashmir, Nepal, Pakistan. *Inula racemosa* Hook. f., is a critically endangered perennial herb distributed throughout the Himalaya; commercially useful in pharmaceutical products.

Morphological characteristics

The species is a perennial herb that grows up to 1.5 metres tall & has a scented root and rootstock. Large number of stems ascend from the rootstock's base. Leaves are leathery, rough above and densely hairy below, 25-50 cm long and 10-12 cm wide, and elliptic-lanceolate in shape, 25-50 cm long and 10-12 cm wide.

Floral biology

Flower heads are yellow with bisexual florets and arise in terminal racemes. Flowering takes place from January to July, while fruiting takes place from October to November. Fruits (achenes) are slender and about 0.5 cm long.

Distribution

In terms of origin and range, it is an oriental species. The species can be found in India's temperate regions of Kashmir, Himachal Pradesh, and Uttarakhand. It is grown in temperate and sub-alpine parts of India, as well as Iran, Afghanistan, Baluchistan, and Europe.

Soil and climate

The cultivation of *Inula* species is best suited to temperate and sub-alpine environments. The crop thrives on well-drained clay-loam soils. It prefers open, sunny environments.

Propagation

Seeds and rootstocks can both be used for propagation. It is, nevertheless, better to develop seedlings from seeds. The fruits (achenes) mature in the winter and can be picked in the high hills by the end of the cold season (March–April).

Inula racemosa produced through micropropagation

Micropropagation has also been proven to be effective. MS (Murashige and Skoog, 1962) medium reinforced with 1.00 mg l-1BA (benzyl adenine), which has been documented as the best performing medium for in vitro shoot multiplication, can be retrieved and multiplied normally. Platelets are toughened and effectively developed in the experimental fields settings at an elevation of roughly 1275 metres above mean sea level after the conserved shoots are retrieved from vitrification.

Nursery technique

Raising propagules

The crop can be cultivated in a nursery by broadcasting seeds in well-prepared nursery beds in November or early March. Germination takes roughly 50 days to finish. After about two months of growth, seedlings are transplanted onto the field with a ball of earth.

Propagule rate and pretreatment

For one hectare of land, around 1 kg seeds are required to raise approximately 40, 000 seedlings. Before sowing, no special seed treatment is required.

Land preparation and fertilizer application

To loosen the soil and incorporate organic manure or FYM, the field should be lightly ploughed. Typically, FYM is applied at a rate of 15 tonnes per hectare during land preparation. When using inorganic fertilisers, half of the nitrogen

(100 kg/hectare) and the whole dose of phosphorus and potassium (100 kg/hectare and 50 kg/hectare, respectively) are applied during land preparation.

Transplanting and optimum spacing

After roughly two months, the germinated seedlings can be transplanted to the main field. In the main field, 50 cm × 50 cm is the optimum spacing. The above-mentioned plant density allows for an optimal crop stand of around 40, 000 plants per hectare.

Intercropping system

Puskarmool is preferred to be cultivated as a sole crop without intercropping.

Interculture and maintenance practices

To get a high biomass output, NPK (nitrogen, phosphorus, potassium) at 200 kg/hectare, 100 kg/hectare, and 50 kg/hectare, respectively, may be employed per year. Six to eight weeks after transplanting, nitrogen is frequently administered in two split doses, one as a basal dose and the other (100 kg/hectare) as a top dressing. During the crop's establishing phase, only manual weeding is suggested twice.

Irrigation practices

Light irrigation every three to four weeks is sufficient to keep the crop healthy.

Disease and pest control

No serious incidence of disease or pest has been observed in this crop.

Crop maturity and harvesting

In places having moderate climate, the crop is harvested in October–November after around a year and a half. After soaking the earth, the roots are dug. The roots have to be thoroughly cleaned and soil particles have to be eliminated.

Post-harvest management

The harvested crop should be cut or sliced into small pieces, dried, and stored in airtight containers. The rootstock and roots should be separated from the aerial portion, chopped into small pieces, and allowed to dry in shade.

Chemical constituents

The rootstock and roots contain inulin, alantolactone, β-sitosterol, isoalantolactone, dihydroalantolactone, and its glucosides. Four sesquiterpene lactones have also been isolated from *Inula* sp.

Roots of *Inula racemosa*

Yield

The crop attains full bloom vested at one-and-a-half years of growth. About 80 quintals of dried roots are obtained from 1 hectare cropped area.

Active principles

The essential oil from the roots is strongly aromatic and contains approximately 60% sesquiterpenes, 22% aplotaxene and 2% phenylacetonitrile Alantolactone, β-sislosterol, daucosteral and its glucosides have been determined in alcoholic extracts of the roots.

Therapeutic uses

Rheumatism, hypertension, cardiovascular and liver disease, respiratory tract issue, lung infections, skin diseases, gastrointestinal disorders, fever, and pain are all treated with the pounded root. It has antispasmodic, antiseptic, anti-inflammatory, antiasthmatic, anthelmintic, diuretic, and antipyretic properties. Chronic bronchitis and rheumatism are treated with this herb. Loss of appetite and gastrointestinal problems are treated with dried rhizomes and roots. The roots are commonly employed as an expectorant and a tonic in indigenous medicine and veterinary care. In many areas, it has also been introduced as a decorative plant and a medicinal herb. The roots are commonly utilised in traditional medicine due to their high quantity of active compounds.

It also has hypoglycemic properties. It's utilised as a stomachic and tonic in veterinary medicine. Seeds have a bitter flavour and are utilised as an aphrodisiac. Leaves and stems are used as fodder and fuel wood, respectively, and its blossoms are employed in religious rites as offerings to numerous deities. People also stuff pieces of root between their garments to keep worms at bay.

20

Common Juniper
Juniperus communis Linn

Cupressaceae

Ayurvedic name	:	Vapusha, Hapusa
Unani name	:	Abhal
Hindi name	:	Aaraar, haubera, abhal
English name	:	Common Juniper
Trade name	:	Hauber
Parts used	:	Berries, leaf, bark and wood

Leaves of *J.communis*

Berries of *J.communis*

Juniperus communis, the **common juniper**, is a species of small tree or shrub in the cypress family Cupressaceae. This evergreen conifer has the largest geographical range of any woody plant, with a circumpolar distribution throughout the cool temperate Northern Hemisphere from the Arctic south in mountains to around 30°N latitude in North America, Europe and Asia. Relict populations can be found in the Atlas Mountains of Africa.

As to be expected from the wide range, *J. communis* is very variable, with several infraspecific taxa; delimitation between the taxa is still uncertain, with genetic data not matching morphological data well.

- Subsp. *communis* – Common juniper. Usually an erect shrub or small tree; leaves long, 8–20(–27) mm; cones small, 5–8 mm, usually shorter than the leaves; found at low to moderate altitude in temperate climates.
- Subsp. *communis* var. *communis* – Europe, most of northern Asia
- Subsp. *communis* var. *depressa* Pursh – North America, Sierra Nevada in California

- Subsp. *communis* var. *hemisphaerica* (J.Presl & C.Presl) Parl. – Mediterranean mountains
- Subsp. *communis* var. *nipponica* (Maxim.) E.H.Wilson – Japan (status uncertain, often treated as *J. rigida* var. *nipponica*)
- Subsp. alpina (Suter) Èelak. – alpine juniper (syn. *J.c.* subsp. *nana, J.c.* var. *saxatilis Pallas, J. sibirica Burgsd.)*. Usually a prostrate ground-hugging shrub; leaves short, 3–8 mm; cones often larger, 7–12 mm, usually longer than the leaves; found in subarctic areas and high altitude alpine zones in temperate areas.
- Subsp. *alpina* var. *alpina* – Greenland, Europe and Asia
- Subsp. *alpina* var. *megistocarpa* Fernald & H.St.John – Eastern Canada (doubtfully distinct from var. *alpina*)
- Subsp. *alpina* var. *jackii* Rehder – Western North America (doubtfully distinct from var. *alpina*)

Morphological characteristics

It is a dense sub-erect or prostate perennial shrub growing up to 1.5 m height. Leaves are 5-13 mm long, in whorls, linear, sharply pointed, spreading nearly at right angles from the branchlets, convex on the back, concave and glaucous bluish white on the upper surface, jointed at base and continued down the stem with a large gland on the decurrent portion.

Fruits *of Juniperus communis*

Floral characteristics

Prostrate specimens of *Juniperus communis* subsp. *alpina*

Flowers are dioecious, axillary, supported by small imbricating bracts. The male cones ovoid, yellow; antheriferous, scales broad ovate and acuminate. Stamens are decussate or in three, connective enlarged ovate or peltate at the apex bearing 2-6 globose pollen sacs near the base. Female flowers are in cones, composed of 2-6 opposite or ternate scales, usually not all fertile. Ovules are 1-2 to each fertile scale Fruit are 7.5-10.0 mm long, sub-globose, blue-black, glaucous berries tips of the scales visible at apex. Seeds are 1-3 with a thick hard testa and often connate into a hard several-celled mass.

Distribution

Juniperus is a temperate species, which occurs in forests throughout Western Himalaya from Srinagar to Kumaon at 1700-4200 m MSL.

Climate and soil

Sandy clay loam soil rich in humus is more suitable for this crop. It prefers temperate climate.

Propagation material

Plant can be multiplied through stem cuttings and seeds, however, meristems and stem-cuttings are ideal material as plants propagated from stem-cuttings grow faster than that of seedlinggs.

Nursery technique

Raising propagules

The right time for collection of seeds is October for raising seedlings of *J. communis*. Moreover, the rate of success of nursery using seeds is very poor. Hence, to multiply the plant vegetative propagation is preferable. Stem cutting is suitable in May. Apical shoot (15cm) performs better root formation than the older part of the stem. Sand is most suitable medium for planting. In propagation chamber or polytunnel, the success rate of rooting is 40 to 57%. IBA 2500 ppm promotes rooting in maximum cuttings (80%) followed by IBA 3000 ppm (60%). Peeling of bark at the lower end of the cuttings improves the rate of success as compared to round or normal cut and crushing at the lower end. Normally, stem-cuttings take 60-70 days for rooting. The rooted stemcuttings should be hardened at least for 6 months before transplanting in field. In nursery, the stem-cuttings should be planted at 10cm X10cm spacing.

Propagule rate and pre-treatment

About 10,000 saplings are required for one hectare land.

Land preparation and manure application

July-August is the suitable period for planting rooted stem-cuttings or seedlings of plant in main field. It is not necessary to plough entire field for planting crop. The planting spots marked at desired space should be made weed free before making planting pits of dimension 45cm X 45cm X 45cm. Pits are filled with equal proportion of soil and sand. About 15 t/ha of Farm Yard Manure (FYM) should be applied as basal dose at the time of transplanting in main the field. In subsequent years, FYM @ 15 t/ha should be applied at the time of bud break in March-April, after winter dormancy.

Transplanting and optimum spacing

Sapling is planted at the centre of the pit and the soil is gently compressed after planting. The pits are made at a spacing of 1m x 1m accommodating 10,000 plants/ha. Plants should be watered lightly after the planting.

Intercropping system

Intercropping is not suitable for the crop.

Irrigation practices

In the absence of rainfall, irrigation at weekly interval is essential for about a month after transplanting to ensure establishment of the saplings in field.

Weed control

Hand weeding twice at monthly intervals during July-September provides effective weed control in the initial 2-3 years of the crop.

Disease and pest control

No disease or insect-pest has so far been noticed in this crop.

Crop maturity and harvesting

J. communis bears flowers / fruits after attaining the age of about five years. Fruits ripe during September-October, hence it is the right time to harvest the fruits. Bark and leaf can also be collected during October.

Post-harvest management

Bark, leaf and fruits can be air dried and stored at room temperature in a dry place.

Chemical constituents

Plant contains 4-terpineol (18.14%), marpol (7.96%), α-pinene (6.96%), ãterpinene (4.46%), β-fenchyl alcohol (1.53%) and oplopenone (0.69%) are major constituents. As per literature, sabinene (48.8%), α-pinene (6.2%) and endofenchyl acetate (5.8%) are major components of the essential oil from needles of *J.communis* growing in nature. Fruit yields 0.8 – 1.2% essential oil and 8% resin. It also contains the bitter substance juniperin.

Male cones of *Juniperus communis*

Yield

Yield could not be recorded because of destructive methods for extracting leaf, bark and wood.

Therapeutic uses

Perfectly shaped *Juniperous communis*

Berries, wood and oil are used in folk remedies for cancer, indurations, polyps, swellings, tumors and warts. Its fruit and essential oil possess carminative, stimulant, deobstruent, diaphoretic, digestive and diuretic properties. They are useful in different forms of dropsies, either administered alone, or in combination with other diuretics. They are also used to treat mucous discharges in gonorrhoea, gleet and leucorrhoea; and some cutaneous diseases. The wood is sudorific in action.

Juniper berries have long been used as medicine by many cultures including the Navajo people. Western American tribes combined the berries of *Juniperus communis* with *Berberis* root bark in a herbal tea. Native Americans also used juniper berries as a female contraceptive.

21

Spikenard *Nardostachys grandiflora* DC; Syn. *Nardostachys jatamansi* DC

Valerianaceae

Ayurvedic name	:	Jatamansi
Unani name	:	Sumbul-ut-teeb
Hindi name	:	Jatamansi
Trade name	:	Jatamansi, Spikenard
Parts used	:	Rhizomes and roots

Nardostachys grandiflora - closer view

Morphological characteristics

Nardostachys species is a perennial rooted herb with a strong woody main root that grows 10-60 cm tall. The rootstock is strong and lengthy, with petioles of withered leaves clinging to it. The stems are pubescent above and glabrate below. Leaves are both Cauline and radical. Cauline leaves are sessile, occur in two to three pairs, are 2.0 cm 0.6 cm in size, decrease in size from base to top, and are oblong or sub-ovate in shape. Radical leaves are large (15-22 cm 1.5-2.5 cm), six to eight in number, longitudinally nerved, slightly pubescent, and narrow down into the petiole, whereas cauline leaves are sessile, occur in two to three pairs, are longitudinally nerved, The jatamansi rhizome is dark brown, tapered, and densely coated with the remains of old leaf bases, giving it the appearance of a heavy beard, and hence the name jatamansi. The fibrous rhizome may be 6-9 cm long with about 20-30 cm-long yellowish taproot. A single root may bear 30-50 rhizomes.

Floral characteristics

Flowering shoots are 30–40 cm tall and produce 40–50 blooms in three to seven condensed cymes at the summit. Flowers are campanulate and range in colour from pinkish red to bluish white. The hairy corolla tube is 6 mm long. The 4-mm-long hairs on the fruit are capped by the sharp, often dentate, calyx teeth.

Flowering takes place between June and August, depending on elevation, while fruits takes place between September and October. Seeds are gathered in October when they turn pale brown in colour.

Distribution

The species can be found in the Himalayas' montane and sub-alpine zones, from Kumaon through Sikkim, Nepal, and Bhutan, at elevations ranging from 3000 to 5000 metres in steep open regions. It is more common in the alpines on western faces, on damp rocky and undisturbed slopes, or on stones with coarse sandy loam soils, and occurs in random patterns.

Climate and soil

The plant is more common on open, stony, and grassy slopes and grows in steep hills with 40°–70° inclined slopes. It is usually found growing on the cooler western slopes.

It is a pioneer species on basic litter and soil deposits. The optimal soil for growing *Nardostachys jatamansi* is loamy permeable soil rich with organic materials like humus. Rapid growth and branching are aided by a thick humus layer.

Propagation material

Seeds are the best propagation material, but if sufficient mother plants are available, vegetative propagation can be used to produce a harvest in two to three years. However, at first, the planting stock must be produced only from seeds. When seeds are put in equal amounts of soil, sand, and FYM (farmyard manure)/compost in styrofoam trays, they have an 80% germination rate.

Nursery technique

Raising propagules

Seeds or vegetative rhizomes collected from the mother plants can be used to establish nursery during May. Small plots of 1 m x 1 m or styrofoam trays in the polyhouse can be used to sow the seeds manually. When sown at a depth of 0.5 cm in sandy soil within the polyhouse, seed germination ranges between 74% and 80%, and takes 12–30 days to germinate.

Nardostachys grandiflora - crop view

Propagule rate and pretreatment

For one hectare of land, around 600 g of seeds are needed to raise seedlings for transplanting.

For quick germination/sprouting, GA_3 (gibberellic acid; 100 PPM [parts per million]) and 200 ppm) are treated to both seeds and rhizomes for 48 hours.

Planting in the field

Land preparation and fertilizer application

Prior to the summer season, land is prepared for cultivation by excavating or ploughing well. The soil is properly tilled, and the beds are left open for a week to allow for solarization. Before seedlings or vegetative rhizomes are transplanted, forest leaf litter/compost/FYM is applied to the beds after the site preparation. Raised beds have been proven to be ideal for high-altitude biomass production. For enhanced plant development and biomass production, a minimum of 40 quintals FYM/forest leaf litter is required as a basal dose. If there isn't enough forest leaf litter, only well-decomposed FYM can be employed.

Transplanting and optimum spacing

Around 50–60 days following germination, the seedlings can be transplanted to the main field at a spacing of 20 cm x 20 cm or 20 cm x 30 cm. Manuring should be completed 15 days prior to the start of transplantation.

Hoeing and earthing-up can be done after manuring. For an area of one hectare, 0.2–0.25 million seedlings are required. During plantation and early growth, there is a high rate of mortality (20%–30%). As a result, extra nursery plants might be kept on hand in the second year to fill in any gaps.

Intercropping system

Intercropping procedures have not found suitable due to the harvesting of roots and rhizomes; therefore, intercropping is not suggested.

Interculture operations and maintenance practices

For 1 hectare of land, 60–80 quintals of manure or forest litter are recommended. During the first year, 50% of the manure is applied, and the remaining is applied in two divided doses during the second and third years.

Irrigation practices

Watering should be done on alternate days at lower altitudes (2000 m) at first to ensure adequate roots. During the dry season, watering is done at weekly intervals. To avoid waterlogging, the soil should be kept at a constant humidity level.

Weed control

Manual weeding is carried out (fortnightly) during early growth season, and later at monthly intervals or as and when required to keep the crop weed-free.

Disease and pest control

No diseases, insects, nematodes or physiological disorders have been observed in this crop.

Harvest management

Crop maturity and harvesting

Maturity period of the plant depends on the propagules used. Plants raised through seeds may take three to four years to mature, while the plants raised through splitting of rhizomes mature in two to three years and are ready for harvesting earlier. To obtain good active chemical ingredients, plants must be collected after senescence in October.

Nardostachys grandiflora - tender growth stage

Post-harvest management

After harvesting in October, the roots should be washed and well dried in shade to reduce their moisture content to 8%–10%. Dried material should be filled in jute bags or wooden boxes, which can be stored in dry godowns. During storage, hairs on root separate due to rubbing, and are often used as dhoop.

Chemical constituents

The dried rhizomes are steam-distilled to yield 1.5%–1.9% of a pale yellow essential oil, commercially known as spikenard oil, emitting a pleasant odour.

Yield

At the experimental site at an elevation of 3600 m, the recorded yield was 835 kg/hectare dry roots; plantation was raised through seedlings. At lower altitudes (2200 m), the re- corded yield was 670 kg/hectare dry weight after third year of cultivation; plantation was raised through seedlings. The yield may increase in subsequent years if the plant is not harvested for one more year.

Therapeutic uses

The rhizomes of this are used in different medicinal system including the traditional medicine. It is used as a good stimulant, antispasmodic, tonic, laxative and antiepileptic. Jatamansi has been traditionally used in treatment of wide range of disorders, which include digestive system, circulatory system, nervous system, respiratory system, urinary system, reproductive system and skin diseases. It also shows marked tranquillizing activity, hypotensive, hypolipidemic, hepatoprotective, neuroprotective, anti-ischemic, antiarrhythmic and anticonvulsant activities. The roots and rhizomes of *Nardostachys jatamansi* have been used to treat hysteria, syncope epilepsy, and mental weakness. It also exhibits cardio protective activity and used in the treatment of neural diseases. The essential oil obtained from the roots shows various pharmacological activities including antimicrobial, antifungal, hypotensive, anti-arrhythmic and anticonvulsant activity.

22

Turpeeth
Operculina turpethum (Linn.) Silva-Manso Syn. *Ipomoea turpethum* R. Bv.

Plant of *Operculina turpethum*

Convolvulaceae

Ayurvedic name	:	Trivrit
Unani name	:	Turbud
Hindi name	:	Nishoth
Trade name	:	Turpeeth, Nishoth
Parts used	:	Root bark, root, and stem

Operculina turpethum (OT) Silva Manso, also known as Tihudi/Trivrit due to its triangular stem, is an Ayurvedic herb from India that belongs to the Convolvulaceae family. Indian Jalap/Turpeth root in English, Nisoth/Turpeth/Panila/Pithori in Hindi, Trivrit in Sanskrit, Tegada in Telugu, Trikolpakkonna/Triputa/Sivata in Malayalam, Sigade in Kannada, and Kumbham/Sivatai in Tamil are some of the other names for it. It grows up to 1000 square feet on Indian roadsides and is occasionally cultivated as an adornment in gardens.

Morphological characteristics

The crop is a fragrant perennial creeper with a basic stem that might be triangular or rectangular. The leaves are many and oval in shape, measuring 2 to 5 inches in length. It consists of cylindrical pieces of root and stem, 1.5-15 cm long x 1-5 cm diameter, with the central woody portion removed by splitting the bark on one side, external surface longitudinally furrowed giving the drug a rope-like appearance, odour distinct but unpleasant or musty, taste bland at first, then slightly acrid, fracture short in bark and fibrous in wood, odour distinct but unpleasant or musty. The root portions are cylindrical, twisted, and a dull grey tint on the outside. The flower arrangement is 1 to 4 inches long and includes three to four branches with white blooms. Fruit is round ½ or ¾ inch in diameter, it bears 4 shiny 2 inch long seeds. The plant bears fruits and flowers from March to December.

Growth and distribution

Trivrit comes in two varieties: Shweta or white Turpeth (OT Silva Manso, Synonym. *I. turpethum*) and Krishna or black Turpeth (OT Silva Manso, Synonym. *I. turpethum*) (the botanical name is *Ipomoea petaloidea* Chois). At an altitude of 900 metres, OT grows all throughout India. It is a native herb that can be found in Australia (Northern Territory, Queensland); Africa (Kenya, Tanzania, Mozambique, Zimbabwe; Western Indian Ocean: Madagascar, Mauritius, Reunion); Pacific (Northwestern Pacific: Micronesia); Asia-Tropical (India, Nepal, Pakistan, Sri Lanka, Indochina, Myanmar, Thailand, Indonesia, Malaysia, Papua New Guinea, Philippines); and Asia-Temperate (China: China-Guangdong; Eastern This plant can be found in the tropical regions of India, America, Pakistan, Sri Lanka, China, the Philippines, Bangladesh, Madagascar, and Mauritania at random.

Varieties

Trivrit has two varieties as *Aruna or Shweta* (i.e., having whitish or reddish coloured root) & *Shyama* (i.e., having blackish root). The botanical name of *Aruna or Shweta Trivrit* is *Operculina turpethum* (L.) Silva Manso (syn. *Ipomoea turpethum*), whilethat of *Shyama* is *Ipomoea petaloides* chois.

Climate and soil

Sandy loam to clay loam soil in moist tropical regions is suitable for the cultivation of this species. It prefers shady and moist places.

Nursery raising

Best method for raising the plants is to use vegetative cuttings of length 10 cm, which should have two nodes. These cuttings may be planted in polybags or directly in mist chamber for rooting. The potting mixture should contain sand, soil, and FYM (farmyard manure) in equal quantities. Best time for raising as pot plants is March–June in mist chambers.

Vegetative cuttings may directly be planted in the field in July. Planting stock can also be raised through seeds, which may be sown in polybags during April–May immediately after collection. Scarified seed germinates after seven days of sowing in the nursery.

Propagule rate and pretreatment

About 2 kg seeds are required for raising planting stock for 1 hectare plantation. Soaking of seeds in water for 24 hours followed by mechanical scarification by rubbing seed coat with sand paper gives 95% germination.

Land preparation and fertilizer application

Land should be ploughed well followed by harrowing twice and planking for good aeration and drainage, since this species is susceptible to waterlogging. About 2 tonnes of FYM per hectare is mixed in the soil during field preparation in June.

Transplanting and optimum spacing

Rooted cuttings or seedlings are transplanted in the well prepared field at a spacing of 30 cm × 30 cm. Thus, the optimum crop stand per hectare is 110 000. Planting is done on ridges to avoid mortality due to waterlogging.

Intercropping system

Since the species is a shade-loving twiner and needs host for support, it can be planted as an under-crop in tree plantations or near hedges and shrubs.

Interculture and maintenance practices

As already stated, about 2 tonnes of FYM per hectare is mixed in soil during field preparation in June before the onset of rains. No further application of any manure is required. Regular manual weeding is recommended after planting and after every 15-day interval.

Irrigation practices

During rainy season, irrigation is seldom required, but in dry spells, flood irrigation at an interval of five to seven days is given till December–January.

Harvesting

The Crop duration is about 10–12 months. Best time for harvesting the roots, stem, and leaves is March–April when the plants begin to dry. Freshly collected roots should be washed thoroughly with fresh water and dried initially in sun for two to three days, followed by drying in shade for the next 10 days till the moisture content reduces to 8%.

Chemical constituents

A glycosidic resin found in the plant's root bark contains the insoluble glycoside turpethein. There are also a lot of secondary metabolites such saponins, flavonoids, glycosides, and phenolics, as well as some essential oil, glucose, and fructose. Turpethein, glycosides, coumarins, scopoletin, saponins, flavonoids, steroids, and carbohydrates are among the chemical ingredients of OT. It contains a number of phytoconstituents, including as glycosidic resin, coumarin, beta-sitosterol, reducing sugars, and essential oils, all of which are beneficial in the treatment of a variety of maladies and diseases.

Yield

Root yield of about 1500 kg/hectare is obtained after 10 months.

Root bark of *Operculina turpethum*

Therapeutic uses

It is used for a variety of therapeutic applications due to its reputation as a powerful medicinal herb. The root bark and seed of this herb are frequently used in Ayurvedic medicine to treat skin disorders like vitiligo, as well as a variety of diseases like cervical lymphadenitis, fistulas, constipation, chronic gout, fever, bronchitis, ulcers, haemorrhoids, tumours, obesity, jaundice, herpes, and induced lacrimation. Rheumatism, flatulence, paralysis, scorpion stings, and snake bites have all been successfully treated with root powder. The root powder has also been reported to be effective in the treatment of hematemesis, herpes, and tuberculosis, while fresh leaf juice is used to cure corneal opacity and conjunctivitis. Avipattikar Churna is an important Ayurvedic formulation of Trivrit (OT) which is used predominantly in disorders pertaining of stomach and intestines.

In the Ayurvedic system, the crop is included in the "ten purgative herbs" group, which supports a therapeutic enema, the "ten antidote herbs" group of herbs that eliminate toxins, and the "colon cleanser, anticancer, and antidote herbs" group. The young leaves of plant and stem are used as a vegetable. The fresh root of the bark is rubbed up with milk and used as a purgative.

23

Indian Trumpet Flower *Oroxylum indicum* Vent.

Bignoniaceae

Pods of *Oroxylum indicum*

Ayurvedic name	:	Shyonak
Unani name	:	Sonapatha
Hindi name	:	Sonapatha
Trade name	:	Sonapatha
Parts used	:	Root bark, stem bark, fruits, seeds, and leaves. Root bark is recommended for medicinal use, but market samples generally contain stem bark.

Oroxylum indicum (L.) Vent. (family Bignoniaceae Juss.) also known as Broken bones tree, Indian trumpet flower, Midnight horror, Shivnak, Sonapatha and Shyonaka. *O. indicum* is being used as medicinal herb for thousands of years without any known adverse effects. The tree is named as "Tree of Damocles" after an incident depicted in an ancient story by Cicero. *O. indicum* Vent is a small or medium sized deciduous tree growing throughout Asia continent. It has been used traditionally for a wide diversity of ailments such as cancer, diarrhea, fever, ulcer, jaundice and arthritis. Destructive and non-sustainable collection methods coupled with low regeneration and habitat destruction have posed serious threats to the survival and availability of this highly useful tree. The existence of *O. indicum* in natural population is highly threatened and it has been categorized as endangered and vulnerable by the Government of India. There have been number of scientific studies conducted to evaluate the toxic effects of the plant. Almost all the studies have shown that *O. indicum* is not toxic to humans even up to high doses.

Morphological characteristics

O. indicum is a small to medium-sized deciduous tree with soft light brown or greyish brown bark and corky lenticels that grows up to 12 metres tall. The leaves are big, 90-180 cm long, 2-3 pinnate with 5 or more pairs of primary pinnae, rachis is fast, cylindrical, swelling at branch junctions, leaflets are 2-4 pairs, oblong or elliptic, acuminate, glabrous. The enormous leaf stalks wither and fall off the tree, accumulating towards the trunk's base and resembling a jumble of broken limb bones.

Inflorescence of *Oroxylum indicum*

Floral characteristics

Flowers are reddish purple on the outside and pale pinkish-yellow on the inside, and they bloom in enormous tall racemes. Bats are attracted to the flowcrs because they bloom at night and generate a strong, foul odour. Long, curving fruit pods dangle from the tree's branches, resembling the wings of a huge bird or dangling sickles or swords in the night. Fruits are sword-shaped flat capsules that are 0.33-1 m long and 5-10 cm wide. The seeds flutter to the earth as the pod breaks open, frequently reaching a long distance and like butterflies. Except at the base, the seeds are numerous, flat, and winged all around like papery wings. In June-July, the plant blooms, and in November, it yields fruit. The root bark is soft and delicious when it is fresh, it is sweet, becoming bitter later. On drying, the bark shrinks, adhere closely to the wood and becomes faintly fissured.

Distribution and relationships

Oroxylum indicum is native to the Indian subcontinent, the Himalayan foothills with a part extending to Bhutan and southern China, Indochina and the Malaysia regions. It is found up to an altitude of 1200 m mainly in ravines, in damp region and moist places in the forests. In India, it is distributed in Himalayan foothills, Eastern and Western Ghats and North East India. It is mostly sighted along the river banks or slopes of the hills. *O. indicum* lives in relationship with the actinomycete *Pseudonocardia oroxyli* present in the soil surrounding the roots.

Soil and Climate

The plant prefers tropical climates with evenly distributed rainfall, between 85 and 130 cm per year. The plant prefers sandy-loam rich soil for growth and development, although it may also thrive in medium to deep black soils to sandy loam soils.

Propagation

The optimum propagation material is seed, which should be harvested before the pods break in February–March. Without any pretreatment, seed germination rates range from 80% to 90%.

Nursery technique

Raising propagules

Before field preparation, a nursery of the plants is raised in polythene bags during second half of March for seedling establishment. The polybags should be filled with sandy-loam soil with good quality, well-decomposed FYM (farmyard manure) added to it in the ratio 2:1. After emergence of the seedlings, the plants are watered regularly to maintain optimum moisture level.

Propagule rate and pretreatment

In all, 250 g of seeds are sufficient to raise stock for plantation on 1 hectare of land. The seeds should be soaked in water for at least 12 hours before sowing to ensure good germination. *O. indicum* is propagated naturally by seeds, which germinate in the beginning of the rainy season (June to September). Seedlings require moderate shade in the early stages. However, the seed set is poor and seed viability is low. Problems related with its natural propagation and indiscriminate exploitation for medicinal purpose has pushed *O. indicum* to the list of endangered plant species of India.

Planting

Land preparation and fertilizer application

The land is tilled to make it porous, friable, and weed-free. Pits of appropriate size, preferably 60 cm × 60 cm × 60 cm, are dug in a square planting geometry (2 m × 2 m). Approximately, 10 kg FYM, 150 g of nitrogen, and 250 g of single super phosphate are mixed in the topsoil, and the pit is loosely filled back before planting.

Transplanting and optimum spacing

Transplanting in the main field is done in the first week of July (with the onset of monsoon in central India). A spacing of 2 m × 2 m is considered appropriate. An optimum crop stand of 2500 plants/hectare is achieved with this spacing. Gap filling is done in September.

Intercropping system

The plant can be grown as a sole crop as well as in a mixed cropping pattern. Short-term seasonal herbs can be grown in the inter-spaces after first year of growth.

Interculture and maintenance practices

About 20 kg of well-decomposed FYM, 150 g of nitrogen, and 250 g of super phosphate are required per plant in three split doses at intervals of six months for two years after transplanting. Potash is required only in potassium-deficient soils. Normally, two weedings along with hoeing are considered sufficient. However, the pit area around the plants may be given more frequent manual weeding, if required.

Irrigation

In the first year, irrigation should be done at least six to eight times but more frequent irrigation during the summer months (at an interval of 7–10 days) is desirable for maximum productivity. The check basin method is found best for irrigation.

Pest and disease

No diseases and nematode attacks have been reported. However, leaf-eating caterpillars like Bihar hairy caterpillar (*Diacrisia obliqua*) and grasshoppers damage leaves during rainy season. Besides, termite attack causes serious damage to root bark in crops grown on drier lands. These insects can be controlled by applying Endosulphan 30 EC spray @ 0.03% in water solution twice at 15-day interval as soon as the termites attack the plants.

Chemical constituents

The chemical constituents obtained from different parts of plant include baicalein-7-O-diglucoside (Oroxylin B), baicalein-7-O-glucoside, chrysin, apegenin, prunetin, sitosterol, oroxindin, biochanin-A, ellagic acid, baicalein and its 6- and 7-glucuronides, scutellarein, tetuin, antraquinone and aloe-emodin.

Harvesting

The tree starts flowering and fruiting after three years, and forms viable seeds even in first fruiting. It takes about three to five months for fruits to attain maturity. The plants are uprooted for harvesting of root bark only after six to eight years, October– December.

Post-harvest management

A pit is dug around the tree and filled with water to facilitate uprooting. After uprooting, the roots are thoroughly cleaned. Thereafter, the bark is peeled off and the root is cut into small pieces, which are shade-dried to reduce their moisture level to less than 12%. The dried material is stored in moisture-proof bags.

Yield

A six-year-old tree yields only 1 kg of dried root bark. Therefore, the estimated yield of root bark is about 4.0 quintals/hectare per year from sixth year onwards..

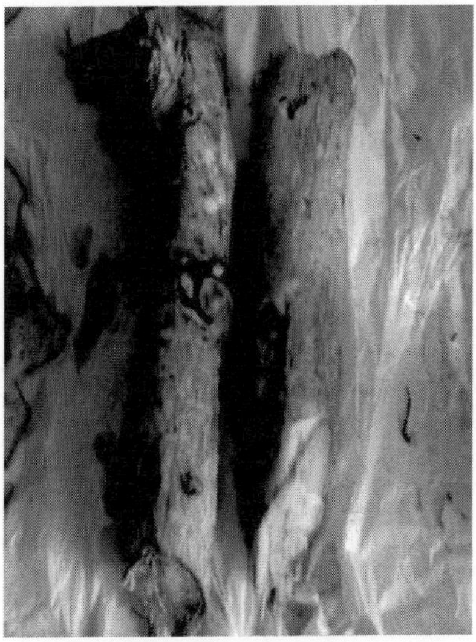

Root bark of *Oroxylum indicum*

Therapeutic uses

Sweet, astringent, bitter, acrid, anti-inflammatory, anodyne, aphrodisiac, expectorant, appetiser, carminative, digestive, anthelmintic, constipating, diaphoretic, diuretic, antiarthritic, antidiabetic, and febrifuges are all properties of roots. Dropsy, cough, sprains, neuralgia, hiccough, asthma, bronchitis, anorexia, dyspepsia, flatulence, colic, diarrhoea, dysentery, strangury, gout, vomiting, leucoderma, wounds, rheumatoid arthritis, and fever are all conditions that tonic can help with. Stomatitis, nasopharyngeal carcinoma, and tuberculosis are all treated with root bark. Stomachic, carminative, and flatulent properties are all employed in the leaves. Rheumatoid arthritis, enlarged spleen, ulcer, cough, and bronchitis are all treated with a leaf decoction. Fruits that have reached maturity are acrid, sweet, anthelmintic, and stomachic. Pharyngodynia, cardiac diseases, gastropathy, bronchitis, haemorrhoids, cough, piles, jaundice, dyspepsia, smallpox, leucoderma, and cholera are all conditions that they can help with. Purgatives are made from seeds. Women use dried seed powder to help them conceive. The seeds produce a non-drying oil that is utilised in the perfume business. For rapid treatment of tonsil pain, the seeds are crushed with fire soot and put to the neck as a paste. The seeds are utilised in Ayurvedic treatment in India, and are found in well-known tonics like Chyawanprash. A decoction of the bark is used to treat gastric ulcers, and a paste of the bark powder is used to treat mouth cancer, scabies, and other skin conditions. Otitis media is relieved by instilling medicinal oil of *O. indicum* in sesame oil basis into the ears.

24

Pistanchio Tree
Pistacia integerrima Stew. ex Brandis

Pistaceaceae

Ayurvedic name	:	Karkarashringi, Shringi
Unani name	:	Kaakarasingi
Hindi name	:	Kakrashinghi
English name	:	Pistanchio tree or Zebra wood
Trade name	:	Kakrashinghi
Parts used	:	Galls, kernels, foliage, bark

Pistacia integerrima - the tree

Pistachia integerrima

Pistacia integerrima Stewart (Pistaceaceae) is an ethnobotanically important tree native to Asia. ***Pistacia integerrima*** is a species of pistachio tree native to Asia, commonly called zebrawood or the crab's claw. It is often classified as Pistacia chinensis ssp. *integerrima*.

Pistacia integerrima, commonly known as kakra shingi, is a moderate-sized deciduous tree, up to 18 m in height, with a short stout bole found in the Himalayas from Indus to Kumaon. Hard, horn-shaped, rugose, hollow galls like excrescences are formed on the leaves and petioles of the plant by an insect of *Pemphigus* species. Dry powdered galls have a very astringent and slightly bitter taste and terebinthine odour. This tree is also used as a rootstock in the cultivation of commercial pistachios

Morphological characteristics

It is a moderate sized deciduous tree with rough grey bark. Leaves are 15-23 cm long with or without a terminal leaflet. Leaflet is stalked, 4-5 sub-opposite pairs, lanceolate, coriaceous, entire and arched. Flowers are greenish-yellow or brownish in colour, dioecious, 0.2 cm diameter, reddish in lateral, puberulous panicles, appearing with or just before the young leaves.

Floral characteristics

The male flowers are in compact panicles, pubescent, 5-15 cm long. Stamens are 5-7 in number. Female flowers lack panicles, 15-26 cm long, elongate, sepals 4, linear and bracts deciduous. Fruit is a drupe, broader than long, glabrous, rugose, grey. Seeds are collected during May-June

Distribution

This plant is distributed over dry scrub forests in North - Western Himalayas between 300 and 2400m MSL. Plant is also cultivated in Punjab.

Climate and Soil

Plant grows well in open rocky grounds and over limestone soil with good amount of organic matter.

Propagation

The crop is propagated through the healthy seeds after careful scarification.

Nursery Technique

Raising Propagules

Seeds are raised in nursery during June - July. About 80-100 kg of seeds is required for one hectare area. Seed germination is only 50%. Seeds are germinated in about 30 days after per cent sowing.

Propagule Rate and Pretreatment

80-100 kg seeds/ha at a spacing of 1m X 1m is required. No pretreatment is required.

Planting in the field

Land preparation and manure application

Land is ploughed once with harrowing and 2-3 ploughing to have a fine tilth. Organic compost (FYM) @ 25 to 30 t/ha is thoroughly mixed with the soil during land preparation.

Transplanting and optimum spacing

Seedlings are transplanted with the onset of monsoon rains during July-August. An optimum spacing of 1m X 1m is given for accommodating 10,000 trees/ha.

Intercropping system

The plant can be grown as a pure crop or herbaceous medicinal plants can be grown as filler crop with it.

Interculture and maintenance practices

Regular weeding and hoeing operations are required after 30 to 60 days of transplantation. Weeding is required for 2-3 years in early stages.

Irrigation practices

It requires irrigation in first year depending upon soil and climatic conditions. Afterwards 2-3 irrigations are required during summer upto 3 years old crop.

Crop maturity and harvesting

The tree has a long span of life, that is 30-40 years. It takes about 9-10 years to bear flowers and fruits in July-August and insect galls are formed in natural habitat after 15-20 years during December-February.

Post-harvest management

Mature galls are hand picked and stored in air-dried places.

Chemical constituents

Essential oil from galls contains apinene, camphene, dl-limonene, 1,8 cineole, α-terpineol, aromadendrene and caprylic acid. Other constituents of galls are β-sitosterol and triterpene acids, pistacienoic acids A and B.

Galls of *Pistachia integerrima*

Yield

An average tree of 15-20 years old produce 0.40 kg of galls. It is a natural phenomenon and no efforts have been made to introduce gall formation to increase yield or its size under plantation. A yield of 25 kg gall may be estimated per ha of plantation.

Therapeutic uses

It is used for a variety of purposes in India, including timber, dye, and fodder. This plant mainly contains alkaloids, flavonoids, tannins, saponins and sterols in different parts including leaf, stem, bark, galls and fruit. Traditionally plant parts particularly its galls have been utilized for treatment of cough, asthma, dysentery, liver disorders and for snake bite. Long, horn-shaped leaf galls that often develop on this tree are harvested and used to make *kakadshringi*, an herbal medicine for diarrhea in northern India.. The galls are aromatic, astringent and expectorant and are valued in Indian Medicine as a remedy for asthma, phthisis and other ailments for the respiratory tract; they are also useful in dysentery, chronic bronchitis, hiccough, vomiting of children, skin diseases, psoriasis, fever, to increase appetite and to remove bed humors. Sushruta prescribed the galls in combination with other drugs for the treatment of snake bite and scorpion sting. *P. integerrima* has depressant action on the central nervous system, analgesic and anti-inflammatory activities and hyperuricemia effect. Galls also possess antiasthmatic, astringent and expectorant properties. Essential oil from galls is antibacterial, antiprotozoal, anthelmintic and antimicrobial in action

25

Indian Podophyllum
Podophyllum hexandrum Royle

Podophyllaceae

Ayurvedic name : Vanyakarkati

Hindi name : Papra, Bankakri

English name : Indian Podophyllum

Trade name : Bankakri

Parts used : Rootstock

Flowers of *Podophyllum hexandrum*

Plants containing lignans have been utilised as folk medicines in traditional medicine of many different cultures for over 1000 years. Plants high in lignin, such as *Kadsura coccinea* (Schizandraceae), *Fraxinus* sp., and *Olea europaea*, have long been employed in Chinese, Japanese, and Eastern world folk medicine. Lignans are an important family of starting products for the development of new therapeutic agents based on structural modifications of such compounds due to their widespread use in traditional medicine. *Podophyllum hexandrum*, a plant with a high lignin concentration, has been employed throughout history and in modern times. It has been used extensively in ancient Ayurvedic medicine to treat a variety of disorders such as Condyloma acuminate, Taenia capitis, monocytoid leukemia, Hodgkin's disease, non-Hodgkin's Lymphoma, cancer of brain, lung, bladder and venereal warts.

The Himalayan locals have long used Indian Podophyllum, with an aqueous extract of the roots being a frequent cathartic. It has also been used as an ophthalmia treatment. Thomson tested the resin from the Indian plant in 1890 and found 56 percent podophyllotoxin concentration. Podwyssotzki discovered podophyllotoxin as the active principle of podophyllin in 1880 and obtained it in a pure state. Several lignans have been found in the rhizomes of *Podophyllum hexandrum*, which are dimerization products of phenylpropanoid pathway intermediates linked by the central carbons of their side chains.

Podophyllum hexandrum is a plant that grows in the Podophyllum genus. In ancient times, Royle (Himalayan Mayapple) was known as Aindri (a divine drug). In Hindi and Ayurveda, it is known as bantrapushi or Giriparpat. This perennial herb, sometimes known as Himalayan May apple or Indian May apple, is found at lower elevations in Himalayan countries such as Afghanistan, Pakistan, India, Nepal, Bhutan, and Southwestern China.

Fruits of *Podophyllum hexandrum*

Distribution

The crop is indigenous to India's Himalayan area (Uttarkashi). It grows as an undergrowth with other herbs at elevations of 2800-3000 m in the Indian Himalayas, in wet alpine meadows, humus-rich and shaded locales, or near stream banks. In India, the Alpine Himalayas (3000-4000 msl) of Jammu and Kashmir, Himachal Pradesh, Sikkim, Uttaranchal, and Arunachal Pradesh are home to *Podophyllum hexandrum*. It has been employed in Kashmir's traditional medicine system since time immemorial, and it is locally known as Banwangun because its red-colored fruit (berry) is the size of a little brinjal.

Morphological characteristics

It has a pale pink flower and vivid red-orange bulbous fruit, and it grows low to the ground with glossy green, lobed leaves on its few rigid branches. It can withstand freezing temperatures, as one would anticipate from a Himalayan plant, but not dry circumstances.

Leaves of *Podophyllum hexandrum*

With deeply lobed leaves and thick stems that rise straight up from the soil, *Podophyllum hexandrum* grows from 12 to 18 inches tall. The name Podophyllum comes from the words pods, which means foot, and phyllon, which means leaf, and refers to the leaves' similarity to a duck's foot. The leaves of the plant are separated into three lobes. They are dark green with brown splotches and fully unfurl once the plant has blossomed. Leaves are alternating, palmate, up to 25 cm across, deeply split into 3-5 lobes, toothed, and purple spotted. White or pale pink 6-petaled flowers appear in the spring at the terminals of strong stalks, followed by plump, oval red berries. The flower, which blooms from May to August, features six petals and six stamens, hence the species name hexandrum, which means "six stamens." The leaves are spherical in shape, 10-25 cm long, and deeply divided into three ovate, toothed lobes, which are sometimes lobed further. The fruit is a huge crimson or reddish berry with many seeds embedded in the pulp, about 2.5-5 cm in diameter. 4–5 percent podophyllum resin is found in American Podophyllum (*P. peltatum*), but 7–16 percent is found in Indian sps. Seasonal changes, varied growth locales, and plant age 21 all contribute to the fluctuation in resin percentage. Resin concentrations of up to 20% have been found in some places. During the flowering period in May-June, the maximum percentage of resin is obtained. As a result, when Indian podophyllum is harvested at the right time, it contains 2.5 times more nutrients.

Climate and soil

The species thrives best as undergrowth as well as in forests in well drained humus rich soil in temperate and subalpine zones.

Propagation

The plant is propagated by seeds and rootstocks.

Nursery raising

Raising propagules

It is propagated from seeds or by division of rootstock. Seed germination is poor. Seeds are to be sown before the onset of winter i.e. in spring season. The germination percentage of seeds ranges from 41-45%. Seedlings exhibit extremely slow growth rate and even after two seasons of growth. They could attain a height of 6.5-7.2 cm. The survival rate of this plants ranges from 85.0-92.6% by this method.

Propagule rate and pretreatment

Seeds of 7.0-8.0 kg seeds are required for an area of one hectare land. No pretreatment of seed is required. A density of 9 plants sq.m. is optimum for first two years.

Land Preparation and manure application

Land should be ploughed and planked to have a fine tilth and make it weed free. Well-decomposed Farm Yard Manure (FYM) @ of 10 t/ha is to be mixed thoroughly with the soil as basal dose.

Transplanting and optimum spacing

Seedlings of 10-12 cm height are transplanted at a spacing of 9m X 9m. It takes about 15 days for establishment.

Interculture operations

Weeding-cum-hoeing may be carried out regularly after 4 weeks interval. Crop normally prefers cool, shady environments and light irrigation during hot season in June-August.

Weed control

Regular weeding, hoeing operations are required during the months from March to May when the crop is grown at lower altitude.

Disease and pest control

No disease or pests are noticed in this crop.

Crop maturity and harvesting

It is a perennial crop. The plants remain in vegetative stage during the first year and flowering start from 2 nd and 3rd year onwards. Rootstocks are harvested in spring season. Rootstocks are found to be rich in resin contents during this season.

Chemical composition

Epipodophyllotoxin, podophyllotoxone, aryltetrahydronaphthalene lignans, flavonoids such as quercetin, quercetin-3-glycoside, podophyllotoxinglycoside, kaempferol, and kaempferol-3-glucoside have all been found in *Podophyllum hexandrum*. Anti-tumor lignans found in the rhizomes and roots include podophyllotoxin, 4'-demethyl podophyllotoxin, and podophyllotoxin 4-O-glucoside

Podophyllum resin or podophyllin is a resinous combination found in the rhizomes and roots. Lignin glycosides, podophyllotoxin, podophyllic acid, picropodophyllin,

-peltalin, and -peltalin are the main ingredients. Gum, starch, albumin, gallic acid, calcium oxalate, and lignin flavones are also found in the rhizomes.

Post-harvest management

Roots and rhizomes are harvested after drying of above ground parts. Roots and rhizomes are then cut into pieces of 15 to 20 cm long, washed and dried in shade. The dried material should be stored in clean containers or gunny bags.

Yield

The crop gives yield of 3.0 - 4.0 t of dried rootstock and 10 kg of seeds from one hectare after 5 years.

Roots of *Podophyllum hexandrum*

Therapeutic uses

Podophyllotoxin is the most important lignan since it is used to make the anti-cancer medicines etoposide, teniposide, and etophos. Lung and testicular malignancies, as well as some leukemias, have all been treated with these chemicals. In addition, podophyllotoxin is the predecessor to CPH 82, a novel rheumatoid arthritis derivative, as well as additional derivatives for the treatment of psoriasis and malaria. Rhizomes are used for typhoid fever, jaundice, dysentery, chronic hepatitis, scofula, rheumatism, skin diseases, tumerous growth, kidney & bladder problems. Plant is also used for gonorrhoea, and syphilis. The podophyllum is used as a purgative and also for treatment of vaginal warts. Two derivatives of podophyllotoxin, called eloposide and teniposide are employed for treatment of cancers. Root paste is applied on ulcers, cuts and wounds.

26

Malabar Kino Tree
Pterocarpus marsupium Roxb.

Fabaceae

Pterocarpus marsupium Roxb.

Ayurvedic name	:	Bijasar, Asan
Unani name	:	Bijasar
Hindi name	:	Vijaysar, Bijasar
Trade name	:	Bijasar, Vijaysar
Parts used	:	Heartwood, bark, leaves, and gum (kino)

Morphological characteristics

Straight bole, longitudinally fissured bark, imparipinnate and elliptic leaves, fragrant flowers in huge panicles, and winged, flat pods distinguish *Pterocarpus* species in the field. The tree can grow up to 30 metres tall and 2.5 metres wide, with a straight and clean bole. The bark is scaly, tough, and fissured longitudinally. Leaflets are five to seven in number, 8–13 cm long, oblong or elliptic in shape, or rotund in shape, with 15–20 pairs of lateral veins. Kino-gum is a fragrant, brittle, almost black oleo-resin obtained from tree trunks that is angular and glistering and comes in minute flakes..

Floral characteristics

Yellow flowers bloom in panicles that are 1–5 cm long and fragrant. Flat, orbicular, winged pods with a diameter of up to 5 cm. Seeds range from one to three in number, are bony, and have a convex form. Flowering starts in November, and fruits lasts until March.

Distribution

The tree can be found in central and peninsular India, primarily in dry mixed deciduous tropical forests in Gujarat and Madhya Pradesh, as well as sub-Himalayan tracts, at elevations of up to 1000 metres. Natural populations have plummeted, and there are hardly any fragile young saplings in the forest. Due to a lack of autogenic reproduction, this species is endangered.

Climate and soil

The tree is native to the tropics and thrives in direct sunlight with moderate rainfall of 80–200 cm. It thrives in fertile, deep clayey loam soil with excellent drainage. In the summer, it can withstand high heat.

Varieties

No improved varieties are available, in this crop.

Propagation material

The plantations are grown from seeds that have been obtained recently. In April–May, mature fruits are harvested from trees before they fall to the ground.

Raising propagules

In April, a nursery is grown from seeds in partial shade. During the monsoons, the plantlets are kept in the nursery until they are about two months old, after which they are transferred to the pits in the main field. Polybags or baskets can also be used to nurture seedlings. If seedlings are to be kept until the next planting season, root coiling in the plants should be avoided. Because seed viability is quite low (about 1%–2%), this tree has been classified as an endangered species category 'A.' Because seedlings generated from tissue culture die out and have a low survival rate in open fields due to their intolerance of high temperatures, winter is the best time to transplant tissue-culture-raised plants into the field.

Propagule rate and pretreatment

To produce seedlings on 1 hectare of ground for planting at 8 m x 8 m spacing, approximately 100 g of viable seeds are required. To protect seeds from fungal infections, they are treated with Thiram at a rate of 3 g/kg of seeds before sowing. Cutting through their ends and immersing them in water for two days before seeding will speed germination. Seed borers should be kept away from newly gathered seeds.

Fertilizer application and land preparation

In April–May, land is ploughed and harvested to create fine tilth. Pits of the proper size (50 cm x 50 cm) are dug at an 8 m x 8 m interval. About 25 kg FYM (farmyard manure), along with 200 g of nitrogen and 150 g of phosphorus, is mixed with soil of each pit as basal dose. The pits are refilled with this mixture after weathering of soil.

Transplanting and optimum spacing

Transplanting can be done in July–August (monsoon season) when the plants are two months old, or it can be done in June–July the following year. The recommended spacing is 8 m x 8 m, which accommodates roughly 160 plants per hectare. In September, the gaps in the field are filled.

Intercropping system

Intercropping with a variety of species such as medicinal plants and vegetable crops is possible when Bijasar is planted at an 8 m x 8 m spacing. At reduced spacing, the species can also be grown as a pure crop.

Interculture and maintenance practices

For the first three years, FYM at 25 kg per plant, nitrogen at 200 g/plant, and phosphorus at 150 g/plant are required. The fertiliser is sprayed in two parts, one in September and the other in January. Two hand weedings are recommended, the first in August and the second in September.

Irrigation practices

In the first year, irrigation should be done six times (ideally once a month) using a check basin system or by filling the pit basin with water.

Disease and pest control

In mature stems and roots, no major insect pests or diseases have been identified. Seeds, on the other hand, are susceptible to seed borer, which reduces seed viability. To some extent, this can be regulated by proper drying (up to 12 percent moisture) and the use of carbon-di-sulphide in storage. Leaf-eating insects and white grub assault are common in the nursery and early growth stages, and can be managed by four sprays of Endosulphan @ 0.003 percent at fortnightly intervals and application of Phorate 10 G in the root zone, respectively. Seed treatment with Thiram @ 3 g/kg of seed is required to keep the plants disease-free in the nursery and early phases of development in the field.

Crop maturity and harvesting

After 10–15 years, the tree is harvested for heartwood production. Before logging a tree, kino gum is gathered by cutting a hole in the bark and drying it in the shade.

Chemical constituents

Pterostilbene (45%), alkaloids (0.4%), tannins (5%), and protein are all found in the plant. Liquiritigenin, isoliquiritigenin, pterostilbene, pterosupin, epicatechin, catechin, kinotannic acid, kinoin, kino red, -eudesmol, carsupin, marsupial, marsupinol, pentosan, and p-hydroxybenzaldehyde are the primary phytoconstituents.

Yield

After 10–15 years, each mature tree produces roughly 500 kg of dry heartwood. As a result, a yield of 750–800 quintals per hectare is projected. Fractures, constipation, haemorrhages, skin illnesses, depurative, ophthalmology, leprosy, rectalgia, and leucoderma, rheumatoid arthritis, skin diseases, external usage for sores, boils, stomach discomfort, and gastrointestinal issues are some of the uses for the leaves. Bark is used as a diuretic, to treat cholera, dysentery, stomachaches, tongue disorders, urinary complaints, and toothaches, as well as an astringent, to treat gland tumours, urethral discharges, chronic ulcers, and as an abortifacient. Neurological issues are treated by stem cells. Heartwood is used as an astringent, bitter acrid, antiinflammatory, anthelmintic, and anodyne, as well as to control blood sugar levels. Fever, Gum-Kino Diarrhea, Passive Hemorrhages, Leucorrhoea, and Dysentery are all treated with flower.

Different parts of the plant *P. marsupium*. (A). Bark, (B). Leaves, (C). Heartwood, (D). Twig, (E). Small branches.

27

Rhubarb
Rheum emodi wall. ex Meissn.

Polygonaceae

Rheum emodi

Ayurvedic name	: Amlavetasa
Unani name	: Rewand chini
Hindi name	: Rewand chini
Trade name	: Revand chini, Rhubarb
Parts used	: Roots and rhizomes

Morphological characteristics

Rewand chini is a perennial stout herb, 1.5–3.0 m tall. Roots are very stout and thick. Leaves are radical, orbicular, or broadly ovate, very large, and 30–45 cm in diameter with long petioles.

Floral characteristics

Flowers are small, dark purple or pale red in axillary panicles. The plant has three to five years of juvenile phase, followed by reproductive phase. Flowering occurs in June–July. Seeds are collected in late August–September, when they turn dark brown. Seed production potential varies from 300 seeds to 950 seeds per plant.

Distribution

The species occurs in the temperate Himalayas, from Kashmir to Sikkim, at an elevation of 2800–4000 m. It is found in alpine zone on rocky soil, moraines, and crevices, between boulders and near streams in specific pockets. It is sporadically cultivated in the North-eastern states for its leaves, which are used as vegetable.

Climate and soil

Rewand chini plant is restricted to the temperate, sub-alpine, and alpine zones of the Himalayas. Well-drained, porous, humus-rich soil is suitable for its cultivation. It prefers exposed or partially shaded habitat and can be cultivated on altitudes above 1800 m in height.

Propagation material

Both seeds and top rhizome segments from mother plants may be used for propagation. Seeds are collected in September–October. They have high and quick germination rate, and require no specific pretreatment.

Micropropagation

Studies reveal that shoot-tip explants of *Rheum emodi* Wall. have given rise to multiple shoots when cultured on a Murashige and Skoog (1962) medium (MS) with 2.0 mg/l 6- benzylaminopurine (BAP) and 1.0 mg/l indole-3-butyric acid (IBA). Also, shoot buds developed from leaf explants using MS medium with 2.0 mg/l BAP and 0.25 to 1.0 mg/l indole-3-acetic acid (IAA) or IBA. Roots were induced when the resulting shoots were placed on MS medium with 1.0 mg/l IBA. Both regeneration procedures gave rise to healthy plantlets that were established in soil under glasshouse conditions at 80% frequency after hardening phase of two weeks.

In vitro produced plants of *Rheum emodi*

Raising propagules

Seedlings are obtained from a nursery raised in March in open sites or in February in sheltered places. Seeds are sown in lines 15 cm apart. In temperate conditions, seeds are sown in the spring season (March–early April). Germination of the seeds is complete within one month of sowing.

Propagule rate and pretreatment

About 600 g of seeds are required to raise a nursery for planting at a spacing of 50 cm × 50 cm in 1 hectare of land. No specific pretreatment is required for the seeds.

Land preparation and fertilizer application

The land may be ploughed and prepared to fine tilth and made aerable. Forest litter/FYM (farm- yard manure) @ 100 quintals/hectare may be used as a basal dose for soil containing 1%–1.5% of organic carbon at 2200 m, applied at the time of land preparation. At lower altitudes (1800 m), higher dose of manure (about 150–200 quintals/hectare) is desirable for maximizing the yield.

Transplanting and optimum spacing

Transplanting is done in May when the plants are almost three months old. The optimum spacing for the crop is 50 cm × 50 cm, which means a crop stand of approximately 40 000 plants per hectare.

Intercropping system

Since the crop of *R. emodi* occupies maximum soil surface, no intercropping is feasible after second year, when planted at a spacing of 50 cm × 50 cm. However, annual vegetable crops or medicinal and aromatic plants can be grown during the first year. Intercropping with *Rheum moorcroftianum* has been reported to be successful. *R. moorcroftianum* has medicinal properties similar to those of *R. emodi*.

Interculture and maintenance practices

While the entire quantity of FYM is applied as basal dose (10 tonnes/hectare), half of nitrogen and full dose of phosphorus and potash are also applied as a basal dose before transplanting the crop. Rest of nitrogen is applied as top dressing six to eight weeks after transplantation.

Irrigation practices

Irrigation is done immediately after transplantation of propagules. This is followed by light irrigation at four-week intervals. Light sprinkler irrigation during the summer season has proved to be useful for the growth of the crop. It also saves water. However, excessive watering may cause decay of underground rhizome, especially in winter months.

Weed control

Regular weeding/hoeing operations, at an interval of 15–30 days, during the establishment and initial growth phase of the crop as well as during monsoons have shown better results in terms of plant growth.

Disease and pest control

The crop is often infected with *Fusarium sp*. Soil drenching with Carbendazim 50 at a rate of 2 g/litre of water is recommended as a control measure. Insects and aphids, which attack the plant at early stage, may be checked by spraying Ecalux 0.5% twice at an interval of 10–15 days.

Crop maturity and harvesting

Crop raised through seedlings needs four to five years for maturity; however, two to three years are enough when raised through rhizome segments. Flowering takes place after one year in May–June and fruiting follows in September–October. The plants attain maturity after about four growing seasons, and may be harvested after the onset of fruiting in September and October.

Post-harvest management

The root and rhizome are cleaned, thoroughly washed, chopped into small pieces, and dried in shade. The dried drug material is packed in airtight containers.

Chemical constituents

Roots contain emodin, emodin-3-monomethyl ether, chropophenol, alsemodin, rhein, sennoside, engenol, rutin, and glucoglatin.

Yield

The expected yield ranges from 37 quintals/hectare to 66 quintals/hectare at an altitude of 1800 m, and between 65 quintals/hectare and 108 quintals/hectare at an altitude of 2200 m, after three years.

Therapeutic uses

Rheum emodi is a mild purgative, astringent, tonic, laxative, stomachic, and aperient. Powdered rhizomes are sprinkled over ulcers for quick healing. It is also reported to be a potent anti- inflammatory drug used as diuretic, liver stimulant, purgative/cathartic, anticholesterolaemic, antitumour, antiseptic and tonic.

28

Manjit, *Rubia cordifolia* L

Asteraceae

Family	: Rubiaceae
Ayurvedic name	: Manjishtha
Unani name	: Majeeth
Hindi name	: Manjitha, Manjit
Trade name	: Manjit
Parts used	: Roots and whole plant

Rubia cordifolia L- the plant

In the coffee family Rubiaceae, *Rubia cordifolia* L. (RC) is a useful medicinal plant in the Ayurvedic system due to its many pharmacological characteristics. It is also known as common Madder or Indian Madder. It was grown for a red pigment generated from the roots. Rubia was divided into roughly 70 species that were found all over the world, including 36 species and two variants described from China. Because of their high bioactivities, Rubia plant extracts and phytochemicals have gotten a lot of interest.

Morphological characteristics

It is a perennial, prickly climber with a stem, growing up to 12 m long. Leaves are highly variable, ovate lanceolate, 5-7 nerved, 2-10 cm long and 2-5 cm broad, occurring in whorls of 4-6. The stems are slender and quadrangular. The leaves are simple, opposite and stipulate, the stipule is interpetiolar. The petiole is 1.5–3 cm long. The blade is lanceolate, 1.5–5 cm × 0.5–2.5 cm, cordate at base, obtuse acuminate at apex and displays three to five longitudinal nerves. Roots are perennial, long, cylindrical, and rusty brown in colour

Floral characteristics

Flowers are small, yellow, and scaly, and occur in terminal cymes. Bracts are ovate and leafy. Calyx is tubular, less than 1 cm long. Corolla is a greenish, divided to the base, tubular with five lobes, and about 3 mm long. Fruits are 4–6 mm in diameter, globose, smooth, shining, violet or purple black in colour with grey black seeds. Flowering occurs in August–September and fruiting in October–November. Flowers are fragrant, minute, whitish or greenish yellow. During August-October plant carries flower and fruit.

Distribution

It is found throughout India, ascending to an altitude of 3750 m from North-West Himalayas eastwards.

Climate and soil

Rubia cordifolia occurs mostly in loamy soil rich in humus. The rainfall is high in the regions in which this plant grows. The plant is a climber and requires support for growth.

Propagation material

The plant is propagated through seeds and two-node root cuttings. The seeds are collected during December and January. It is preferable to use seeds for large-scale cultivation, considering the cost factor and high rate of germination.

Nursery technique

Raising propagules

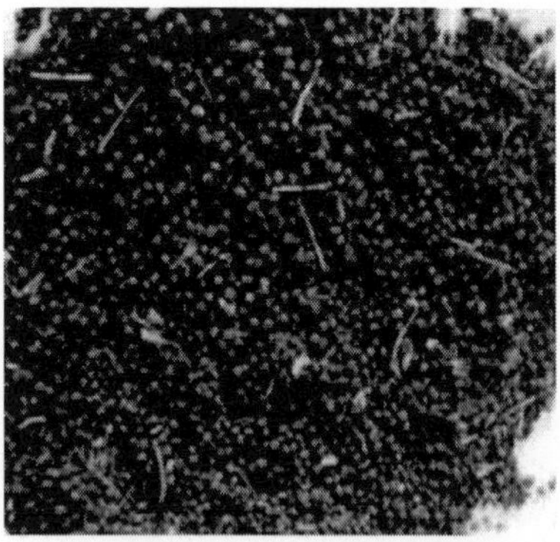

Rubia cordifolia - seeds

The planting stock may be raised in nursery in January through seeds. The seeds obtained from dried ripe black fruits are sown in nursery beds either in rows or randomly by broadcasting. A thin layer of soil and organic manure is spread over the seeds, and the beds are regularly watered. After germination, seedlings with two to three leaves are transplanted in polybags for establishment. The plants can also be raised through cuttings containing two or three nodes, treated with commercially available rooting hormones or 3000 PPM (parts per million) IBA (indole-3-butyric acid) for rooting. Although cuttings give 90% success, it is still preferable to take up large-scale cultivation through seeds, which is economic and results in 80%–85% germination within 20 days.

Propagule rate and pretreatment

About 350 g of seeds are required to raise a nursery for planting in 1 hectare of land. It is advisable to treat the seeds with 0.02% Bavistin 50 WP before sowing. Sometimes, seeds may be directly broadcasted in the well-prepared field or sown in rows. In such cases, much higher quantities of seeds, to the tune of 1.5–2 kg, may be required.

Land preparation and fertilizer application

The soil is properly ploughed, harrowed once or twice, and planked lightly to make it porous and weed- free. FYM (farmyard manure) @ 10 tonnes/hectare and NPK (nitrogen, phosphorus, potassium) @ 30:40:20 kg, respectively, are applied to the soil as basal dose during land preparation.

Transplanting and optimum spacing

Seedlings/rooted cuttings are transplanted to the main field in April– May. An optimum spacing of 60 cm × 75 cm is recommended in the field, which gives an optimum crop stand of 22,000 plants per hectare. When the seeds are directly sown in rows or broadcast in the well-prepared field, singling is an important activity to provide optimum spacing to the growing plants.

Intercropping system

The plant is a climber and may be intercropped with shrubby, perennial species, serving as host/support crop.

Interculture and maintenance practices

FYM @ 5 tonnes/hectare or nitrogen @ 20 kg/hectare is applied 120–130 days after transplanting by top dressing. In the second and third years, NPK may each be applied @ 20 kg/hectare by top dressing with the onset of rainfall after the intercultural operation. In the first year, first manual weeding cum hoeing is done 45 days after planting, and the second one is carried out 120 days after planting. Bamboo/shrub staking is done 30–45 days after transplanting. Intercultural operations during second and third years may be carried out at least twice during each year.

Irrigation practices

Irrigation may be provided as and when necessary. However, the plant may be grown as a rainfed crop in North-East India where the rainfall is high and evenly distributed throughout the year.

Crop maturity and harvesting

R. cordifolia starts flowering after one year during August and seeds mature in October–November. Roots can be harvested after two years at preflowering stage in October or late fruiting stage by the end of November or even in early December when seed is required for next crop. The crop can, however, be allowed to stand in the field for three years.

Rubia cordifolia - dry roots

Post-harvest management

The hard roots are cut into small pieces and dried in the shade. The dried root pieces are packed in gunny bags for storage in cool and dry place.

Chemical constituents

Roots contain resinous and extractive matter, gum, sugar, colouring matter, and salts of lime. The colouring matter consists of purpirin, manjistin, garancin, xanthine, besides several anthraquinone derivates.

Yield

About 3 tonnes of dry root is obtained per hectare area of cultivated crop.

Therapeutic uses

Skin problems and spleen disorders are treated with powdered dried roots and fruits are administered internally. It is used to treat serious burns, ulcers, and bone fractures. It is used as a tonic, antitussive, and in the treatment of chronic low fevers. Internally, the roots are used to treat abnormal uterine bleeding, internal and external haemorrhage, bronchitis, rheumatism, kidney, bladder, and gallstones, dysentery, and other ailments. Blood problems can be treated using this plant. Alterative, anodyne, antiphlogistic, astringent, diuretic, expectorant, styptic, and vulnerary are some therapeutic properties of the roots.

29

Kuchila
Strychnos nux-vomica Linn

Strychnaceae (Loganiaceae)

Ayurvedic name	:	Visamusti
Unani name	:	Azaraqi, Kuchla
Hindi name	:	Kuchla
Trade name	:	Kuchila
Parts used	:	Dried ripe seeds and stem bark

Morphological characteristics

Strychnos nux-vomica is a medium sized deciduous tree. The trunk is tall, thick, straight and cylindrical covered with yellowish-grey to dark grey, smooth and thin barks. Wood is dense, hard white and close grained. Branches are irregular and covered with smooth ash colored bark and shiny dark green young shoots. Leaves are broad, opposite, simple and entire, smooth, short-stalked, blade orbicular to broadly elliptical or ovate, base rounded to cordate, apex shortly acuminate or acute, glabrous and shiny above, minutely hairy especially on veins beneath, 3–5-veined from the base.

Floral characteristics

Inflorescence is terminal cymes, many-flowered and appears with young leaves at the end of branchlets or on axillary shoots. Flowers hermaphrodite, actinomorphic, homogamous, greenish-white colored, small in size, nectariferous, funnel-shaped and emits unpleasant odour. Flower are pentamerous; calyx lobes ovate, outside densely hairy. Corolla bears a slender tube of 1 cm long, abruptly widening at the throat, outside glabrous, having pubescence at base, lobes narrowly ovate, of 3 mm long, margin thickened and minutely hairy, greenish white to white; stamens inserted at the mouth of the corolla tube, exserted altering with corolla lobes. Anthers pale cream in color, dithecous, introrse and dehisce longitudinally. Ovary is superior, ovoid, glabrous, bicarpellary with axile placentation. The style is filiform, glabrous and as long as the corolla tube, stigma capitates (head-shaped).

Fruit is a globose, indehiscent berry with a smooth and hard shell. The fruit is full of soft, white, fleshy and jelly-like pulp containing 1-5 seeds. Seeds are lens-shaped to orbicular or ellipsoid, covered with dense silky hairs radiating from the centre, and giving a characteristic shine to the seed. One side of seed is concave and the other convex with a small depression in the centre of each side. Endosperm is very hard, dark grey in color, odourless and bitter in taste containing small embryo.

Strychnos nux-vomica- fully matured fruits

Strychnos nux-vomica- Inflorescence

Climate and Soil

Kopilu is a deciduous plant of tropical and sub-tropical climate and thrives well in areas receiving 750-2000 mm annual rainfall and mean annual temperature ranging from 25-35^0 C. It grows well on a wide variety of soils, ranging from loamy, loamy-sandy soil and lateritic soils but on poorly drained heavy acidic soils. Soil pH should be slightly shows stunted growth acidic in nature i.e. 5.5 to 6.0.

Propagation

Propagation through seed

Ripe berries are collected from trees during March-April. The seeds are removed from the fleshy pericarp, cleaned, dried and sorted. 750-800 seeds weigh one kilogram. Seed is orthodox and mesobiotic in nature. Seeds are desiccation tolerant, dry seed (10% moisture content) can be hermitically stored at ambient temperature for 30 weeks without losing their viability. Seeds exhibit physiological dormancy and hence slow and erratic germination (10% germination) is recorded from freshly harvested seeds (Sivakumar *et al.*, 2006). Dormancy can be removed (after-ripening) by storing seeds with 10% moisture content at ambient temperature for 30 weeks or by treating with hormones like gibberellic acid (GA3) or indole butyric acid (IBA). Treating seeds with gibberellic acid (Soaking seeds in 500 ppm gibberellic acid (GA3) for 24 h), or incubating seeds at 40 ^{0}C for 3 days or alternate water soaking (16 h) and drying (8 h) for 14 days, the germination can be improved up to 32%. But 92 percentage of seed germination can be achieved by using ripened seeds. Growth of seedlings is very slow but it can be improved by using medium sized polythene bags containing pot mixture of farm yard manure (FYM), soil and sand in ratio of 1:1:1(Chaubey and Krishnamurthy, 2015).

Propagation through vegetative methods

Macro-propagation through cuttings and root suckers is suitable. Semi-hardwood cuttings are prepared in early summer and kept under moist conditions after treating with rooting hormones. Rooting percentage is quite low, often less than 25%. Root suckers from mother plant are pricked out and transplanted in polybags for eventual planting out (Anonymous, 2008; Somashekhar and Sharma, 2002).

Micro-propagation/ clonal propagation of kupilu can be successfully done through shoot proliferation and hypocotyl culture in suitable media. For shoot proliferation, nodal explants excised from healthy mature trees are cultured on Murashige and Skoog medium with various additions. IAA (5 micro M) and NAA (10 micro M) were the most suitable growth regulators, both promoting shoot and root formation (Kumar 1990). For hypocotyl culture seeds were germinated in MS medium supplemented with 1000ppm gibberellic acid (GA^3) under dark condition. Hypocotyls segments of 3mm size are cultured in MS medium with various additions and maintained at 25±10 C with 16 hr illumination at fluorescent light. Kinetin is most suitable for shoot formation and NAA is best for root formation (Kumar and Datta, 1989).

Land preparation and planting

Kopilu is light demanding plant and must be planted in the open field condition. Site should be deeply ploughed till soil achieves fine tilth and weed free. Planting is done in pits. Pit size of 60 cm X 60 cm X 60 cm is dug at a spacing of 5 m X 5 m (400 plants/ha) or 6m X 6m (278 plants/ha) depending upon the rain fall of the area and fertility status of the soil. These pits are filled with mixture of soil, 20 kg of well decomposed cow dung or compost and basal dose of fertilizer (50 g DAP). Chloropyriphos dust (20%EC) @50 grams per pit should be applied if there is a chance of attack of white ant. Proportionate quantity of sand should be mixed if the soil is heavy. One year old seedlings raised in containers are planted in May-June. After planting, proper pressing at the base of the seedling and stacking should be done to avoid bending or breakage by wind. Planting in the evening hours and initial irrigation improves seedling survival percentage.

Intercultural operations

Intercultural operations should be carried out as per the requirement. Annually 2-3 weedings may be carried out for keeping weed free during initial growth periods. After each weeding, soil working should be done along with application of supplementary doses of organic manure and inorganic fertilizers.

Manures and fertilizers

Well decomposed cow dung/ compost @ 20 kg/pit should be mixed with the soil during planting of seedlings. For getting good canopy growth a total of 15-20 kg well decomposed FYM/plant/year for 2-10 old plants should be applied in two split doses. At the time of planting 50g of DAP/plant should be mixed with the soil. At 30 days after planting 50g N/plant should be applied in two split doses. 2nd year onwards 100g NPK per plant should be applied in split doses after cleaning and hoeing the area around basin of the plant. First dose of fertilizer should be applied in June-July and second dose in October –November.

Irrigation

It is drought hardy species but irrigation at 7-10 days intervals helps during initial years is beneficial for seedling height growth. Irrigation after fruit set particularly during pre monsoon drought period is essential for reducing immature fruit drop.

Intercropping

Kopilu is a slow growing plant. Hence during initial 6-8 years intercrops like bottle gourd, lady's finger, cucumber, ridge gourd green gram can be grown successfully in the inter spaces. Intercropping of pulse crops like green gram, black gram etc. is beneficial during early years.

Tending operations

Fruit and seed yield is proportionate to the size of the tree. Coppicing or pollarding of trees reduces the size of the tree and causes a loss in seed production which is never likely to be made up. Pruning has the same effect but to a lesser degree. Coppicing very young trees may result ultimately in acquiring their crowns of fuller spread than if they were left to grow naturally, but trees tend to develop spreading crowns as soon as height growth starts falling off. Artificial measures to promote crown spread are therefore not likely to be worthwhile

Pest and diseases

Plant is normally free of pest and diseases but in few instances leaf spot disease caused by *Colletotrichum gloeosporioides* is being reported. The initial symptoms were small greyish brown spots usually surrounded by a yellow halo. The spots enlarged rapidly, coalesced and often involved large areas of the leaf. Slight blistering of the infected region is evident. With severe spotting the leaf gradually dries up. Infection is restricted to matured leaves only. Young leaves are free of leaf spot disease but on maturation gets infected. The disease can be controlled by clean management practices and spraying with copper or streptomycin based fungicide.

Flowering and fruiting leaf fall occurs in the cold season, mostly in December and January. New shiny leaves unfolds at the advent of summer season (1st week of February), followed by flowering initiation. Blooming progress through March and terminates at the end of April with peak of 30 days between mid February to mid March. Flowers open in the evening hours between 4.00 pm to 6.30 pm. and anther dehisces longitudinally half an hour after anthesis. Flowering continues two to three weeks from the first flower initiation. Heavy flower drop occurs and fruit set percentage depends up the visiting frequency of insect pollinators and only 25-30% fruit set takes place. Heavy young fruit drop is observed in prolonged dry spells. Fruits take about 10-11 months for maturing and attains orange red colour on ripening. Number of seeds per fruit varies one to five but 3-4 are common.

Chemical constituents

Dried seeds of *Strychnos nux-vomica*

Almost all part of this tree contains alkaloids of medicinal importance but seed is official in many pharmacopeia. It is rich in alkaloids, flavonoids, tannins and triterpenoids, glycosides, lignins and steroids. More than 90 chemical compounds have been isolated from different parts of nux vomica but strychnine and brucine are the principal toxic alkaloids. They occur not only in the seeds but also in roots, wood, bark, fruit pulp and hard fruit shells. Seeds contain 2.6 to 3.0% of total alkaloids, of which 1.25 to 2.5% is strychnine and 1.5 to 1.7% is brucine. The seeds additionally contain chlorogenic acid, a glycoside (loganin), and 3.0% of fixed oil

Harvesting and yield

Fruiting in plantation rose from propagules of seed origin starts after 15-20 years of planting. The orange-light red colour of fruit indicates maturity. Fruits are collected December onwards periodically when they mature. Fleshy pericarp is favourite food of many birds, monkeys, squirrels and hence should not be left un-plucked for a long period. A crop stand of 400 trees per hectare and average range of 50-75 kg of dry seeds per tree per year, a yield of 12- 20 t/ha is expected from a 20 year plantation annually. Once a *nux-vomica* plantation rose, profitable yields can be obtained for about 35-40 years. The yield of seed from individual trees fluctuates remarkably from year to year. Drought and fire especially the latter, causes a subsequent drop in yield.

Post-harvest management

Matured fruits are manually collected, fleshy pericarp was removed, washed properly and shade dried up to 10 % moisture content. Seeds are then graded according to size. Seeds of light weight and those float on the surface of water are discarded. Storage does not affect the total alkaloid content of seeds particularly strychnine and brucine. Processed mature seeds can be stored in gunny bags at ambient temperature in the usual way for a long time or till preparation of tincture.

Therapeutic uses

The Ayurvedic Pharmacopoeia of India recommends detoxified seeds for ailment of paralysis, facial paralysis, sciatica and impotency. The root bark is useful in cholera and intermittent fever. The leaves are applied as poultice in the treatment of chronic wounds and ulcers. The pulp of the ripe fruit is used in treating paralytic affections of palms and foot. The seeds are bitter and used as aphrodisiac, appetizer, anti-periodic, digestive, purgative, and stimulant. They are applied in the treatment of anemia, lumbago, asthma, bronchitis, constipation, diabetes, malarial fever, skin disease, paralysis, muscle weakness, emphysema, colic, insomnia, nervous debility, dyspepsia, diarrhoea, hysteria, mental emotions, epilepsy, chronic constipation, gout, chronic rheumatism, hydrophobia and spermatorrhoea. Pharmacologically the drug made from the plant has been tested on animals for its action as antifungal, antibacterial, antioxidant, anti-inflammation, anticancerous, antidiabetic, neuropharmacological properties as well as for the specific actions on gastrointestinal, bone cells and cardiovascular systems. Apart from medicinal use Kupilu have also been used as rodenticide, avicide, insecticide, Nematicide and Piscicide.

30

Rohida Tree

Tecomella undulata (Sm.) Seem. Syn. *Tecoma undulata* (Sm.) G. Don

Bignoniaceae

Tecoma undulata - the tree

Ayurvedic name	:	Rohitak, Rohira
Hindi name	:	Rugtrora, Rohitaka
English name	:	Rohida Tree
Trade name	:	Rakhta rohida (Rohida Tree)
Parts used	:	Bark

Morphological characteristics

Tecomella undulata is a slow growing small deciduous tree with drooping branches and grey-tomentose innovations which are otherwise glabrous. Leaves are usually opposite, 5- 10 cm long, simple, narrowly oblong, obtuse and entire with undulate margins.

Floral characteristics

Flowers are pale-yellow to deep orange, arranged in few flowered corymbose racemes on short lateral branches. Calyx is campanulate, lobes broadly ovate, obtuse and mucronate. Corolla is companulate, veined, stamens exserted, flaments glabrous, stigma 2-lamillate. Capsules are slightly curved, smooth and seeds are winged.

Distribution

Plant is distributed in drier parts of North-West and Western India extending eastwards to the river Yamuna and ascending to an altitude of 1200 meter in the outer Himalaya.

Climate and Soil The plant prefers subtropical climate having medium to light soil.

Propagation material

Seed.

Raising propagules

Seeds are sown in the month of May on flat beds/raised nursery bed or polybags.

Land preparation and fertilizer application

Pits of 60 x 60 x 60 cm are to be prepared at an optimum spacing of 4 meter between row to row and 3 meter between plant to plant and filled with FYM. NPK in the ratio of 750: 600:300 g/plant is given as single basal dose.

Transplanting and optimum spacing

Plantation of seedlings is done at a spacing of 3X4 meter in the month of July-August.

Intercropping system

Aromatic grasses / annual species of medicinal plants can be cultivated as intercrop.

Interculture and maintenance practice

One to two harrowing is to be given in the middle space for keeping the field clean.

Irrigation practices

Irrigation may be given as per the season during winter at 30-40 days interval and summer at 20-30 days interval. After establishment of the crop, it is not necessary to give frequent irrigations.

Weed control

Weeding is done as and when required to keep the field weed free.

Disease and pest control

The plants need to be protected from termites and stem borer for which suitable insecticides are used.

Crop maturity and harvesting

The crop matures after 3-4 years and should be harvested for its bark during summer (April-May).

Post-harvest management

The bark is shade dried and stored in gunny bags in dry, ventilated places for marketing.

Chemical constituents

Bark contains alkanes; heart wood containsiridoid glycosides 6-o-veratroyl catalposide and tecomelloside and flowers contain flavonoids.

Various plant parts of *Tecoma undulata*

Therapeutic uses

Medicinal properties of the plant have been investigated both, in light of recent scientific developments as well as traditionally. The powder of bark is specially recommended in ascites with hepatosplenomegaly. It is an excellent blood purifier hence, rewarding in hepatitis. Bark forms a major constituent of various herbal formulations like Livo-plus, Liv- 52, Herboliv, Amylcure, Livosan, Exol etc for curing inflammatory hepatic diseases. Besides it possesses significant anticancer, immunomodulatory, cytotoxic, analgesic and antibacterial activities. It is also used in the treatment of eczema. The bark of the plant has mild relaxant, cardio tonic, and chloretic (An agent, usually a drug, that stimulates the liver to increase output of bile) activities. The bark obtained from the young branches is often employed as a remedy for syphilis. The bark is powdered and is used with hot milk for few days for abortion. Its seeds are crushed with extract of Pinus leaves, and taken to cure haemorrhoids. *T. undulata* leaves have oleanolic acid, ursolic acid and betulinic acid, compounds that are strong HIV prohibitors. Plant is anti-leucorrhoeic, used in enlargement of spleen and leucoderma; bark febrifuge, antisyphilitic, given in liver diseases. Flowers are used in diabetes.

31

Giloe
Tinospora cordifolia (Willd.) Miers ex HK. f. and Th. Syn. *Meninspermum cordifolium* Willd

Menispermaceae

Ayurvedic name	: Amrita, Guduchi
Unani name	: Giloe
Hindi name	: Giloe, Gurcha
Trade name	: Giloe
Parts used	: Stem, root, whole plant

Leaves of *Tinospora cordifolia*

Tinospora cordifolia (Thunb.) Miers (family: Menispermaceae) commonly known as "Amrita" or" Guduchi" of "Giloy" is a climbing shrub found throughout India. *T. cordifolia* has high medicinal value in the world and is also the number one recommended natural herb for the Indian system of medicine (ISM) which has been used over centuries to treat various diseases. The roots and stems contain several secondary metabolites having curative properties. The species is endemic to India and is common throughout tropical and subtropical zones at an altitude of 600 m.

Species

- *Tinospora cordifolia* (Willd.) Miers ex Hook. F & Thoms
- *Cocculus cordifolius* DC.
- *Menispermum cardifolium* Willd
- *T. glabra* (N. Burm.) Merr.

Morphological characteristics

Fruits of *Tinospora cordifolia*

It is a big climber that grows over the highest trees in the forest and produces aerial roots up to 10 metres long. It's a sprawling climbing shrub with a slew of elongated twining branches. Giloy leaves are simple, exstipulate, alternating, with long petioles up to 15 cm long, pulvinate, and roundish at the base and apex, with the basal one being longer and twisted partially and halfway around. Oval-cordate or roughly oval lamina, 15-20 cm long and 10-15 cm wide, 7 nerved and deeply cordate at base, glabrous above, membranous, whitish yellow with a conspicuous reticulum beneath.

Floral characteristics

T. cordifolia flowers are unisexual, with greenish yellow blooms on axillary and terminal racemes, grouped male flowers, and solitary female flowers. The outer ones are smaller than the inner ones. Sepals 6, free in two sets of three each. Petals have six petals, are smaller than sepals, and are obovate and membranous in shape. Fruits are a cluster of 1-3 ovoid smooth drupelets on a stout stem with crimson or orange-colored subterminal style scars.

Origin and distribution

The crop is indigenous to tropical regions of the Indian subcontinent. The plant distributed throughout the tropical region of India up to 1,200m above sea level from Kumaon to Assam, in north extending through West Bengal, Bihar, Deccan, Kankan, Karnataka and Kerala. It is indigenous to areas of India, Myanmar, Sri Lanka, China, Thailand, Philippines, Indonesia, Malaysia, Borneo, Vietnam, Bangladesh, North Africa, and South Africa. 7-10 It is typically grown in deciduous and dry forests at elevations up to 1000m.

Soil

It grows well in almost any type of soils and under varying climatic conditions. Medium black or red soil with acidic to neutral pH are found to be most suitable for the cultivation of this crop. Well drained and soils rich in organic matter content are ideal.

Climate

It is a very hardy plant which is found growing in varied type of climates in several parts of th country. However, for its luxuriant growth, it prefers warm and humid tropical climate, with an annual rainfall of 2000-3000 mm.

Land preparation

The land should be ploughed 2 to 3 times and leveled properly. Pits of 30 cm^3 should be dug at the recommended spacing and filled with FYM and top soil in equal proportion well in advance of planting season.

Variety:

The variety RRL (J)-82 is found suitable for cultivation and quality yield.

Nursery raising and planting

The plant is cultivated by stem cutting in the month of May-June. It requires some support preferably Neem and Mango trees, such plants are supposed to possess better medicinal values.

Cultivation

Propagation

It is propagated through seeds and cuttings. Seeds have poor germination capacity, therefore for economic cultivation, it is propagated through cuttings. About 10-15 cm long stem cuttings having atleast two nodes are planted in raised beds or in poly bags to get greater success. Treatment of cut ends with rooting hormones gives better results. The rooting of cuttings takes about 8-10 weeks and cuttings will be ready for planting in the main field after about 10 weeks.

Planting

The usual planting time is during the onset of monsoon in May-June. Rooted cuttings are planted in main field at a spacing of 1 m x 1 m. About 10,000 cuttings are required to plant in an hectare area. The pits have to be watered immediately after planting and at regular intervals there after till they are established.

Training

Since *Tinospora* is a climber, proper training structures are necessary to get higher yields. This is achieved by using proper training structures like wires or live supports. *Tinospora* plants when planted in neem plantations are found to be highly compatible with each other. Neem trees of 5-6 year old could be utilized as live supports.

Manures and fertilizers

As organic crop, the plant can be raised using organic manures like, Farm Yard Manure @10t/ ha, Vermi-Compost @2t/ ha and Green Manure @10t/ ha that results in good yield. To prevent diseases, bio-pesticides could be prepared (either single or mixture) from Neem (kernel, seeds & leaves) and Cow's urine etc.

Considering the inorganic fertilizers, Usually it is a practice to apply 10 tonnes of well rotten Farm Yard Manure per hectare. In addition fertilizer dose of 60-40-40 kg N-P-K are applied/ha to obtain good yield. Of the recommended quantity, half of N and full dose of P_2O_5 and K_2O are applied as basal dose at the time of planting and the remaining half of N is applied after two months of planting.

Irrigation

The field after plantation should be irrigated periodically weekly during summer season and fortnightly during winter season

Weeding and hoeing

Periodical hoeing is done, both in the nursery and field as per requirement.

Pests and diseases

Pest and disease problems are low in this crop. However, the sooty mold disease is found to damage the leaves and the crop. The sooty mold fungus grown on gummy excreta of the insects deposited on the leaves. This can be controlled by spraying a solution of Nuvacron.

Harvesting/post harvesting operation

The first harvest of leaves is obtained after 3-4 months of planting and subsequently at two months interval. When the leaves begin to fall in August-September, the stems should be cut 30 cm above the ground and collected. While harvesting, care should be taken to pluck the leaves without damaging the vines. Pale green coloured stems are cut into small pieces and dried in shade.

Chemical composition/ Nutritive value

Tinospora contains diverse array of phytochemicals, including alkaloids, phytosterols, glycosides and mixed other chemical compounds like tinosporaside, jatrorrhizine, palmatine, berberine, tembeterine, tinocordifolioside, phenylpropene disaccharides, tinospora acid, tinospora, tinosporin, and tinosporide have been isolated. The leaves are bitter in taste and astringent and contain furanoid bitter principle tinosporine, tinosporide, tinosporidine, cordifol, heptacosanol and actacosanol as its major constituents. Stems contain β-sitosterol.

T. cordifolia contains with high fibre (15.8%), protein (4.5%-11.2%), carbohydrate (61.66%), and low fat (3.1%). *T. cordifolia* nutritive value is 292.54 calories per 100 grams. It has high potassium (0.845%), high chromium (0.006%), sufficient iron (0.28%) and sufficient calcium (0.131%), important in various regulatory functions.

Yield

An average yield of about 10-15 q./ha of dried leaves can be obtained from one hectare in an year.

Therapeutic uses

This plant has great potential for developing useful drugs. The leaves extract has shown anti-HIV 1 activity. Thus, it can be said that biological extract from this plant will certainly be helpful in protecting and treating various viral diseases in humans. *T. cordifolia* has an importance in traditional ayurvedic medicine used for ages in the treatment of fever, chronic diarrhea, cancer, jaundice, dysentery, bo ne fracture, pain, asthma, skin disease, poisonous insect, snake bite, eye disorders. A dry bark of *T. cordifolia* has anti-spasmodic, antipyretic, antiallergic, anti-inflammatory and anti-leprotic properties.

Extracts of the endophytic fungus *Nigrospora sphaerica* obtained from *T. cordifolia* were found to have insecticidal properties against the Oriental leafworm moth (*Spodoptera litura*), a polyphagous pest.

32

Vellapine
Vateria indica Linn.

Dipterocarpaceae

Ayurvedic name	:	Sarja
Unani name	:	Raal Safed
Hindi name	:	Kahruba, Dammar
English name	:	White Damar, Indian Copal-Tree
Trade name	:	Vellapine, White Dhup, White Damar, Piney resion
Parts used	:	Resin of trunk

Vateria indica- the tree

Morphological characteristics

A large elegant evergreen tree, grows up to 30 m. high, with a clean, cylindrical bole of 15 m. and a girth of 4.5 m. Bark is rough, whitish to grey, peeling off in thick round flakes. Leaves are coriaceous, Ovate to Oblong, entire (Leaf falls in March, new foliage appears in April-May, the second flush of foliage starts after rains, in October to December.) Fruit is a Capsule, ovoid, pale brown, fleshy, 8 to 11cm long, and 3.5 to 6 cm in Diameter. 1 seeded, reddish white, filled with fat. Ripens during June to July.

Floral characteristics

Flowers are bisexual, about 2 cm across, white, slightly fragrant, arranged in panicles. Panicle is robust, multi-branched, up to 15 cm long and drooping. Fruit is capsules, 4-6 cm X 2-4 cm in size, pale brown, fleshy, hard when dry, splitting by 3 valves when ripe.

Control of nursery pests and diseases

Seedlings in the nursery may be slightly infested by the bagworm, *Pteroma plagiophleps* and leaf-webber *Rhodoneura* species, which may not require any control measures.

The weevil *Indomias hispidulus* feeding on tender foliage and termites affecting the roots of nursery seedlings may also occur. Severe incidence of leaf-blight caused by *Altemaria altemata* may be seen in both seed-bed and container seedlings. Application of Carbendazim (@ 0.2% a.i.) as foliar spray can control the disease.

Leaf-spot disease caused by *Cylindrocladium quinqueseptatum. Colletotrichum gloeosporioides, Bipolaris maydis* and *Phoma* species can be controlled using the same fungicide, applied at weekly intervals.

Plantation establishment

Shaded area with moist deciduous or semievergreen forest area is selected to out-plant the seedlings of *V. indica at* a spacing of 2m x 2m, or more by digging pits of 30 cm x 30 cm x 30 cm size for planting the seedlings Weeding is done to remove the undergrowth and ground flora. Poly-potted seedlings are planted and maintained in the shaded nursery by the onset of South-West monsoon in June. Polythene covers are removed without damaging the root system of the seedlings and the cover may be hanged on the stakes which are fixed at each pit-point, while aligning the plot. The seedling should be planted in such a way that the level of ground is tallying with the level of soil around the seedling. Direct sowing of seeds after soil work in the site is also reported to be successful.

Plantation maintenance and growth of seedlings

Almost 90 per cent of the out-planted seedlings survive during the first month of planting. There can be a decline in the survival percentage to 60, mainly due to drought. During the first summer-season after field-planting, mulching and shade are essential to improve the survival and growth of transplanted seedlings. The seedlings grow to an average height of 83 cm within ten months after field-planting.

Control of pests and diseases in plantation

No major pest problem is observed in the field-planted seedlings of *V. indica.* Mild incidence of leaf-spot due to the infection of *Pestalotiopsis* species and *Colletotrichumgloeosporioides* may occur, which may not require any control measure. The felling age of the tree is reported to be 30 years.

Crop maturity and harvesting

The tree becomes ready to yield resin after 20 years of age.

Chemical constituents

The tree yields a resin, which is complex mixture of triterpenes. Seeds yield, a semi solid fat known as piney tallow of Malabar. Gum yield lipids namely Vateriaphenol A, Vateriaphenol B, Vateriaphenol C and Vaticanol C and have been found active against tumours. The bark is reported to be containing Polyphenols – dl – epi-catechin, laevorotatory isomers of fisetinidol, fzelechin and bergenin. The leaves and roots contain bergenin and hopeaphenol. The seeds also contain bergenin. The fruit shell contains 25% tannins. High-performance liquid chromatography analysis showed that the extract contains bergenin, hopeaphenol, vaticanol B, vaticanol C, and å-viniferin.

Yield

Yield not estimated under plantation

Therapeutic uses

Tallow is used as an Emollient, Stimulant, Antibacterial & useful application in chronic rheumatism. Bark is acrid, used as an alexipharmic. Leaf Juice is applied to cure burns and is also used in diseases of blood. Resin is credited with Tonic, Carminative & Expectorant properties and used in several diseases, such as Throat disorders, Cough, Chronic bronchitis, Piles, Diarrhoea, Rheumatism, Tubercular glands, Boils etc.

Climate and soil

The plant grows well in cool and moist climate but heavy and frequent rain is fatal during blooming.Sandy loam soil is best for its cultivation.

Propagation material

The plant is generally propagated through separation of the new plantlets that develop from the runners. As much as 50–60 new plantlets may be separated from one mother plant. This crop is also propagated through seeds that show about 80% germination.

Nursery technique

Raising propagules

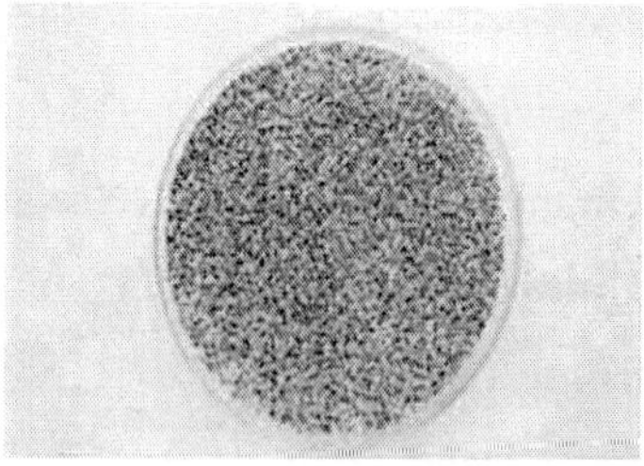

Viola serpens – the seeds

The planting stock can also be easily raised through division of mother plants, which are directly planted in the field. To raise the planting stock through seeds, they are sown in the nursery in March– April, 10–15 cm apart from line to line. About 80% seeds germinate within three weeks. Sowing of the seed directly in the field is not recommended, as the seed size is very small.

Propagule rate and pretreatment

About 110 000 plantlets are required for planting in 1 hectare of land. When seeds are used as propagation material, about 1–1.5 kg seeds per hectare are required. No seed treatment is required before raising the nursery, as the seed coat is not very hard.

Land preparation and fertilizer application

Field should be well prepared through two to three ploughings, followed by planking to make the soil suitable for the transplanting of the plantlets. Bed should be raised or oriented in such a way that there is proper drainage and no water stagnation, especially due to heavy rains in the rainy season. Exposure to frequent and heavy rains is very damaging for the flowers. In places of water scarcity, sunken beds may be prepared to conserve moisture. Well-decomposed FYM (farmyard manure) @ 12 tonnes/hectare should be applied to the soil before transplantation of the seedlings to obtain maximum yield. The seedlings are planted at a spacing of 30 cm × 30 cm.

Transplanting and optimum spacing

The crop is grown in kharif season during May–June, with the onset of pre-monsoon rains. This is the best period for establishment of plantlets in the field. When the crop is propagated through division of the mother plants, the separated plantlets can be planted directly in the main field in May–June. About one-month-old seedlings can be transplanted in the main field, at an optimum spacing of 30 cm × 30 cm.

Intercropping system

The crops can be grown in orchards where partial shade conditions prevail, provided no waterlogging takes place.

Interculture and maintenance practices

Only organic manure @ 12 tonnes/hectare is recommended as a basal dose. The field should be free from weeds, especially at the initial stages of plant establishment. Field should be drained well by digging channels across the field, especially during the rainy season. Irrigation should be provided as and when required during hot weather. For promoting the growth of the plants, 50 PPM (parts per million) solution of GA^3 (gibberellic acid) may be sprayed at an interval of one month from September to May. Manual hoeing and weeding once in a month are recommended to control the weeds.

Disease and pest control

Browning and blotching of the leaves with dead areas having distinct black margins may occur during rains. These infected leaves should be collected and either burnt or buried deep in the soil to check the further spread of the disease. Pesticides or fungicides should never be used.

Crop maturity and harvesting

Leaves can be harvested after two months of establishment of the crop and subsequent harvesting may be done at an interval of one month till December. The plant starts flowering after 9–10 months of growth during February–March, corresponding with the increase in temperature. Fruits can be harvested in April and May. Flowers and seeds should be harvested either in February–March or April–May, depending on the climatic conditions of the area.

Post-harvest management

Flowers and seeds should be air-dried in shade and packed in polythene bags. The flowers should be packed in airtight polythene bags and stored in dark place at room temperature. The seeds retain viability for long periods when stored in airtight containers.

Chemical constituents

Voiline, ionone, and ionine are the main active principles of the plant. Roots, leaves, and blossoms also contain methyl-salicylate in the form of a glucoside. Leaves of Viola contain rutin content in the range 0.15% – 0.45%.

Therapeutic uses

Viola species is antipyretic, diaphioretic, diuretic, aperient, antipyretic, and febrifuge in action. It is useful in respiratory track congestion, asthma, sore throat, cold, coryza, and cancer of the throat. It is also very useful in bleeding piles, constipation, fever, headache, and skin diseases. In the Unani system, this plant is the main ingredient of Joshanda, consisting of a mixture of drugs, used mainly for cough and cold in the form of a decoction.

Colour Plates

Chapter 2: Devil's Cotton*Abroma augusta* L.

Abroma augusta- the plant

Abroma augusta- the flower

Abroma augusta- the seeds

Chapter 3: Indian Aconite *Aconitum balfourii* Stapf

Inflorescence of *A. balfourii*

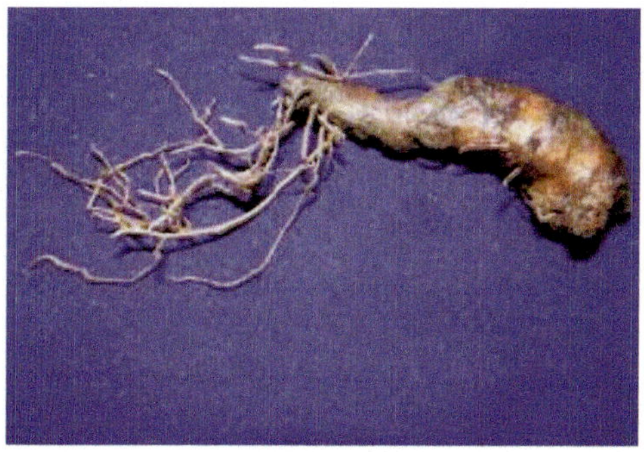

Tubers of *A. balfourii*

Chapter 4: Angelica *Angelica glauca* Edgew

Angelica glauca

Nursery technique

Seedlings of *Angelica glauca*

Fruits

Rootstock

Chapter 5: Aloewood *Aquilaria malaccensis* Lam. syn. *A. agallocha* Roxb.

Aquilaria malaccensis

Chapter 6: Danti *Baliospermum montanum* Muell. Arg

Baliospermum montanum - the plant

Baliospermum montanum - seeds

Baliospermum montanum –the roots

Chapter 7: Brazil Wood, *Caesalpinia sappan* L.

Caesalpinia sappan - the tree

Callicarpa macrophylla – the plant

Caesalpinia sappan - the flower

Caesalpinia sappan - the seeds

Chapter 8: Perfumed Cherry *Callicarpa macrophylla* Vahl.

Callicarpa macrophylla – the plant

Chapter 9: Gandrayan *Cinnamomum camphora* Nees & Eberm

Camphor tree

Camphor powder

Chapter 10: Tree TurmericCoscinium fenestratum (Gaertn.) Colebr.

Coscinium fenestratum – the plant

Fruits and seeds of *Coscinium fenestratum* (Gaertn.) Colebr

Chapter 11: Kalimusali *Curculigo orchioides* Gaertn

Field view of *Curculigo orchioides*

Cut rhizome pieces of *Curculigo orchioides*

Chapter 12: Black Zedoary *Curcuma caesia* Roxb

Plants of *Curcuma caesia*

Flower and rhizome of *Curcuma caesia*

Chapter 13: White turmeric *Curcuma zedoaria* (Christ.) Rosc

Plant of *Curcuma zedoaria*

Flowers of *Curcuma zedoaria*

Rhizomes of *Curcuma zedoaria*

Chapter 14: Himalayan Marsh Orchid *Dactylorhiza hatagirea* (D.Don) Soo

Inflorescence of *Dactylorhiza hatagirea*

Acclimatization of collected saplings in the glass house

Tubers of *Dactylorhiza hatagirea*

Chapter 15: Rattalu, *Dioscorea bulbifera* L

Leaf, Tuber and the Whole plant of *D. bulbifera*

Chapter 16: Shankhapuspi *Evolvulus alsinoides* Linn.

Evolvulus alsinoides - the plant

Chapter 17: White TeakGmelina arborea Roxb

Seedlings of *Gmelina arborea*

Leaves of *Gmelina arborea*

Bark of *Gmelina arborea*

Chapter 18: Gurmar *Gymnema sylvestre* R. Br. ex Schult.

Chapter 19: Puskarmool *Inula racemosa* Hook. F

Inula racemosa - field view

Inula racemosa produced through micropropagation

Roots of *Inula racemosa*

Chapter 20: Common Juniper *Juniperus communis* Linn

Leaves of *J.communis*

Berries of *J.communis*

Fruits *of Juniperus communis*

Prostrate specimens of *Juniperus communis* subsp. *alpina*

Male cones of *Juniperus communis*

Perfectly shaped *Juniperous communis*

Chapter 21: Spikenard *Nardostachys grandiflora* DC; Syn. *Nardostachys jatamansi* DC

Nardostachys grandiflora - closer view

Nardostachys grandiflora - crop view

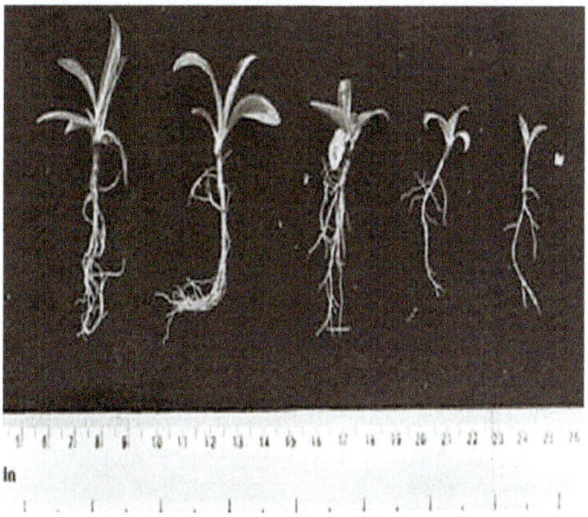

Nardostachys grandiflora - tender growth stage

Chapter 22: Turpeeth *Operculina turpethum* (Linn.) Silva-Manso Syn. *Ipomoea turpethum* R. Bv.

Plant of *Operculina turpethum*

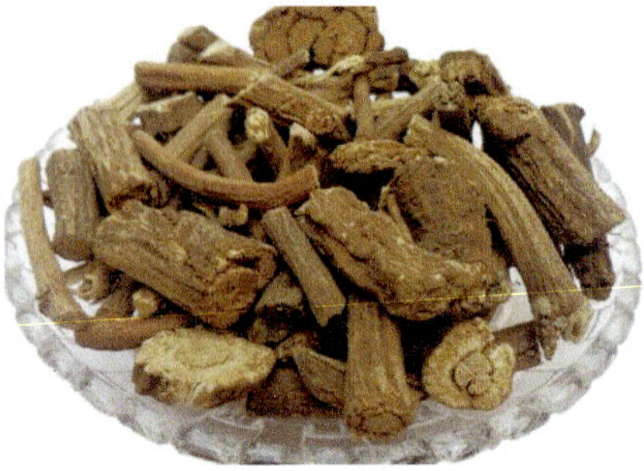

Root bark of *Operculina turpethum*

Chapter 23: Indian Trumpet Flower *Oroxylum indicum* Vent.

Pods of *Oroxylum indicum*

Inflorescence of *Oroxylum indicum*

Root bark of *Oroxylum indicum*

Chapter 24: Pistanchio Tree *Pistacia integerrima* Stew. ex Brandis

Pistacia integerrima - the tree

Pistachia integerrima

Galls of *Pistachia integerrima*

Chapter 25: Indian Podophyllum *Podophyllum hexandrum* Royle

Flowers of *Podophyllum hexandrum*

Fruits of *Podophyllum hexandrum*

Leaves of *Podophyllum hexandrum*

Roots of *Podophyllum hexandrum*

Chapter 26: Malabar Kino Tree *Pterocarpus marsupium* Roxb.

Different parts of the plant *P. marsupium*. (A). Bark, (B). Leaves, (C). Heartwood, (D). Twig, (E). Small branches.

Chapter 27: Malabar Kino Tree *Pterocarpus marsupium* Roxb.

Rheum emodi

In vitro produced plants of *Rheum emodi*

Chapter 28: Manjit, ***Rubia cordifolia*** L

Rubia cordifolia L- the plant

Rubia cordifolia - seeds

Rubia cordifolia - dry roots

Chapter 29: Kuchila *Strychnos nux-vomica* Linn

Strychnos nux-vomica- fully matured fruits

Strychnos nux-vomica- Inflorescence

Dried seeds of *Strychnos nux-vomica*

Chapter 30: Rohida Tree *Tecomella undulata* (Sm.) Seem. Syn. *Tecoma undulata* (Sm.) G. Don

Tecoma undulata - the tree

Various plant parts of *Tecoma undulata*

Chapter 31: Giloe *Tinospora cordifolia* (Willd.) Miers ex HK. f. and Th. Syn. *Meninspermum cordifolium* Willd

Leaves of *Tinospora cordifolia*

Fruits of *Tinospora cordifolia*

Chapter 32: Vellapine *Vateria indica* Linn.

Vateria indica- the tree

Chapter 33: Banafsha *Viola serpens* Linn.

Viola serpens – the plant

Viola serpens – the seeds

References

Adams, R.P. (2004). Junipers of the World: The genus *Juniperus. Victoria: Trafford.* 4120-4250.

Airi, S., Rawal, R.S., Dhar U. and Purohit, A.N. (1997). Population studies on *Podophyllum hexandrum* Royle- A dwindling, medicinal plant of the Himalaya. *Plant Genet. Resources*, 110: 29-34.

Anonymous (2008). Agro-techniques of selected medicinal plants. National Medicinal Plants Board, Department of AYUSH, Ministry of Health and Family Welfare, Government of India. 2: 179-182.

Ansari S.H., Ansari, M. Ali, J.S. Qadry (1993). Essential oils *of Pistacia integerrima* galls and their effect on the central nervous system. *International Journal of Pharmacognosy,* 31(2): 89-95.

Bhattacharjee, S.K. (2005). Use of flavours and fragrances. In: Bhattacharje S.K., editor. *Handbook of aromatic plants*. 2nd ed. Jaipur: Pointer Publishers.

Bishwa Bhusan Kundua, Karan Vannia, Ayesha Farheena, Priyanka JHAC, Devendra Kumar and Vijay Kumar (2020). *Dioscorea bulbifera* L. (Dioscoreaceae): A review of its ethnobotany, pharmacology and conservation needs. *South African Journal of Botany*; 110.

Burke, J.P., Williams, K., Narayan K.M., Leibson C., Hafener S.M., Stern M.P. (2003). A population perspective on diabetes prevention: whom should we target for preventing weight gain. Diabetes Care. 26(7): 1999–2004.

Chauhan, N.S. (1999). Medicinal and Aromatic plants of Himachal Pradesh. 1st ed. New Delhi: Indus Publishing; *Oroxylum indicum;* pp. 96-298.

Choudhary, D.K., Kaul, B.L. and Khan S. (1998). Cultivation and conservation of *Podophyllum hexandrum*- An overview. *J. Med. Aro. Plant Sci.*; 20: 1071-1073.

Dalal, N.V. and Rai V.R. (2004). *In vitro propagation of Oroxylum indicum* Vent. *A medicinally important forest tree. J. For. Res*. 29: 61-5.

Duke, J.A. and Judith L., Ducellier (1993). Handbook of Alternative Cash Crops. *CRC Press.*

F. Austin (1982). *Operculina turpethum* as a Medicinal Plant in Asia. Eco. *Bot.*, 260: 265.

Giri, A. and Arasu, M.L. (2000). Transgenic hairy roots: recent trends and applications. *Biotechnology Advances.*, 18: 1-22.

Grover J.K., Yadav S. and Vats, V. (2002). Medicinal plants of India with anti-diabetic potential. *Journal of Ethnopharmacology*. 81(1): 81-100.

Gupta, R. and Sethi, K.L. (1983). Conservation of medicinal plant resources. In: Jain S.K. and Mehra K.L. (eds) Conservation of tropical plant resources, 101-107.

http://nurserylive.com

Jackson, D.E., Dewick, P.M. (1984). *Aryltetralin lignans* from *Podophyllum hexandrum* and *Podophyllum peltatum*. *Phytochem.*, 23(5): 1147-1152.

Jayaram, K. and Prasad, M.N. (2008). *Genetic diversity in Oroxylum indicum (L.) Vent. (Bignoniaceae), a vulnerable medicinal plant by random amplified polymorphic* DNA *marker. Afr.J.Biotech.* 7: 254-62.

K.P. Srikantha Murty, Bhavprakasha and Bhavmishra (2008). Vol. I, Chaukhamba Shri Krishna Das, Varanasi, India, 258: 259.

Kamil M.W. and Dewick, P.M. (1986). Biosynthetic relationship of aryltetralin lactone lignans to dibenzylbutyrolactone lignans. *Phytochem*. 25: 2093-2102.

Khare, C.P. (2004). *Oroxylum indicum.* Indian Herbal Remedies: Rational Western Therapy, Ayurvedic and other traditional usage, Botany. *4th ed. New York: Springer-Verlag Berlin Heidelberg*; pp. 340.

Kumar A. and Datta, S.K. (1989). Plantlet regeneration from hypocotyl tissue of *Strychnos nux-vomica* Linn. *Current Science*, 58(14): 812-813.

Kumar, A. (1990). Regeneration of forest tree *Strychnos nux vomica* L. through tissue culture. *Indian Journal of Forestry*, 13(3): 263-265.

Kumar, S., Singh, J., Shah N.C. and Ranjan, V. (1997). Indian Medicinal Plants Facing Genetic Erosion. *Central Institute of Medicinal & Aromatic Plants, Lucknow*; pp 205.

Manohar S.H., Naik P.M., Praveen N., Murthy H.N. (2009). Distribution of gymnemic acid in various organs of *Gymnema sylvestre*. *Journal of Forestry Research*, 20(3): 268-270.

Nadkarni, K.M. and Nadkarni A.K. (1999). *Indian Materia Medica* Vol. I, *Popular Prakashan Ltd. Bombay*, 131: 965.

Nand Lal and Paramvir Singh Ahuja (1989). Propagation of Indian Rhubarb (*Rheum emodi* Wall.) using shoot-tip and leaf explant culture, *Plant Cell Reports,* 8: 493-496.

Nayar, S.L. and Chopra, I.C. (1951). Distribution of British pharmacopoeial drug plants and their substitutes growing in India. *Pharmaceutical and Drugs Research Committee, Council of Scientific and Industrial Research, New Delhi*, 56 pp.

Nicholson, J.W. (1937). Results of coppicing, pollarding and pruning experiments to stimulate *Strychnos Nux-vomica* fruit production. The Indian Forester, 63(9): 588-597.

Pant, S. and S.S. Samant (2010). Ethnobotanical observations in the Mornaula Reserve Forest of Kumoun, West Himalaya, India. *Ethnobotanical Leaflets*, pp .193.

Pistacia integerrima (2014). Germplasm Resources Information Network (GRIN). Agricultural Research Service (ARS), United States Department of Agriculture (USDA). (second ed.), Academic Publishers, Delhi. pp. 377.

Pieroni and Andrea (2005). *Prance, Ghillean; Nesbitt, Mark (eds.). The Cultural History of Plants. Routledge.* p. 36.

Prasanthi, Donipati, S., Hara Sreeramulu (2015). Preliminary Phytochemical Screening of *Curcuma caesia. International Journal of Current Microbiology and Applied Sciences*, 4(11): 30-34.

Punkambekar, S.V. (1947). Effect of storage on the alkaloidal content of *Strychnos nuxvomica* seeds. *Current science,* 16(11): 346.

R.N. Chopra, I.C. Chopra, K.L. Handa, L.D. Kapoor (1982). Indigenous Drugs of India.

Rajeshwari, Sahu and Jyoti Saxena (2013). A brief review on medicinal value of *Curcuma caecia.* I*nternational journal of pharmacy and life sciences,* 5(4): 2664-2666.

Sankaran, K.V., Rahman, M.U. aand Zachariah, S. (1980). Leaf spot disease of *Strychnos nuxvomica* caused by *Colletotrichchum gloeosporioides* in South India. *Plant Disease*, 64: 953-954.

Schmelzer, G.H. and Gurib-Fakim M.A. (2008). Plant Resources of Tropical Africa (PROTA), Medicinal Plants, Backhuys Publishers, Wageningen, Netherlands. 11(1): 575-577.

Selvam, A.B.D. (2012). Pharmacognosy of Negative Listed Plants. Pp. 37-47.

Shane M.C. and Whorter L. (2009). Dietary supplements for diabetes: an evaluation of commonly used products. *Diabetes Spectrum*. 22(4): 206-213.

Sharma and A.K. Gaur (2012). *Aconitum balfourii* Stapf: A rare medicinal herb from Alpine., *Himalayan Journal of Medicinal Plants Research*, 6(22): 3810-3817.

Shreekar Pant and Tsewang Rinchen (2012). *Dactylorhiza hatagirea*: A high value medicinal orchid. *Journal of Medicinal Plants Research*, 6(19): 3522-3524.

Singh V.K., Umar S., Ansari A. and Iqbal M. (2008). *Gymnema sylvestre* for diabetics. *Journal of Herbs, Spices and Medicinal Plants*, 14 (1-2: 88–106).

Sivakumar, V., Anandalakshmi R., Warrier, R.R., Tigabu, M., Oden, P.C., Vijaya Chandran S.N., Geetha S. and Singh, B.G. (2006). Effects of presowing treatments, desiccation and storage conditions on germination of *Strychnosnux-vomica* seeds, a valuable medicinal plant. New Forests, 32: 121-131.

Smith, C.M. and Reynard, A.M. (1995). Essentials of Pharmacology. *Philadelphia, Pa, USA: WB Saunders.*

Somashekhar, B.S. and Sharma, M. (2002). Training manual on propagation techniques of commercially important medicinal plants. Andhra Pradesh State Forest Department. Pp. 110-111.

Sun, W., Wang S., Zhao W., Wu C., Guo S., Gao H., Tao H., Lu J., Wang and Chen Y. (2017). Chemical constituents and biological research on plants in the genus Curcuma. *Critical Reviews in Food Science and Nutrition*, 57(7): 1451-1523.

Ullah H.M., Arif, Zaman, Sayera, Juhara, Fatematuj, Akter, Lucky, Tareq, Syed Mohammed, Masum, Emranul Haque and Bhattacharjee, Rajib (2014). Evaluation of antinociceptive, in-vivo & in-vitro anti-inflammatory activity of ethanolic extract of *Curcuma zedoaria* rhizome". *BMC Complementary and Alternative Medicine.* 14: 346.

Uniyal, M.R. (1977). Uttarakhand Vanaushdh Darshika. Pub. *Indian system of Medicine and Homeopathy,* New Delhi.

Warrier, P.K., Nambiar V.P., Ramankutty and Vasudevan R. (1995). *editors.* Indian Medicinal Plants: A compendium of 500 species. *1st ed. Chennai: O*rient Longmam Private Ltd; *Oroxylum indicum*; pp. 186-190.

Yoganarasimhan, S.N. (1996). V*ol. 1. Bangalore: Interline Publishing, Medicinal plants* of India: Karnataka; pp. 366.